ETHICAL CITIES

Combining elements of sustainable and resilient cities agendas, together with those from social justice studies, and incorporating concerns about good governance, transparency and accountability, the book presents a coherent conceptual framework for the ethical city, in which to embed existing and new activities within cities so as to guide local action.

The authors' observations are derived from city-specific surveys and urban case studies. These reveal how progressive cities are promoting a diverse range of ethically informed approaches to urbanism, such as community wealth building, basic income initiatives, participatory budgeting and citizen assemblies. The text argues that the ethical city is a logical next step for critical urbanism in the era of late capitalism, characterised by divisive politics, burgeoning inequality, widespread technology-induced disruptions to every aspect of modern life and existential threats posed by climate change, sustainability imperatives and pandemics. Engaging with their communities in meaningful ways and promoting positive transformative change, ethical cities are well placed to deliver liveable and sustainable places for all, rather than only for wealthy elites. Likewise, the aftermath of shocks such as the 2008 Global Financial Crisis and the Covid-19 pandemic reveals that cities that are not purposeful in addressing inequalities, social problems, unsustainability and corruption face deepening difficulties.

Readers from across physical and social sciences, humanities and arts, as well as across policy, business and civil society, will find that the application of ethical principles is key to the pursuit of socially inclusive urban futures and the potential for cities and their communities to emerge from or, at least, ameliorate a diverse range of local, national and global challenges.

Brendan F.D. Barrett (FRSA) is Specially Appointed Professor in the Center for the Study of Co★ Design, Osaka University, Japan. He has worked in academia and environmental consultancy in the UK, Japan and Australia and as an official with UNEP and the United Nations University.

Ralph Horne is Professor of Geography and Deputy Pro-Vice Chancellor, Research and Innovation for the College of Design and Social Context at RMIT University, Australia. He combines research leadership and participation in research projects concerning the environmental, social and policy context of production and consumption in the urban environment.

John Fien is Professor of Practice in the School of Architecture and Urban Design at RMIT University, Australia. With an academic career spanning four decades, he was previously Executive Director and Professor at the Swinburne Leadership Institute, Professor of Sustainability at RMIT University and Professor of Environmental Education at Griffith University.

'The attainment of sustainable urban futures is an imperative for humankind as ever more people will live in cities. Critical urban studies aim to achieve this goal. Barrett, Horne and Fien develop a strong case for an ethical framework to help cities re-orientate and more effectively respond to the major challenges facing the world today, including evolving patterns of economic competition, technological disruptions, climate change and pandemics. Examining and drawing insights from contemporary initiatives in cities across the globe, the book represents an important contribution to critical urban studies. It will be greatly appreciated by both urban practitioners and scholars.'

Hans van Ginkel, *Former Rector and Emeritus Professor of Geography, Utrecht University, Former United Nations Under-Secretary-General and Rector of the United Nations University*

'This engaging book dives into the duality of cities as both collective and individualistic spaces. Drawing upon numerous examples from cities across the globe, it presents pathways for collective action that would help mitigate contemporary urban and global problems, including poverty and the climate crisis. The authors, though not shying away from the hard reality of life in many cities, present an optimistic perspective in the framing of an ethical city – one that is both deliberate and inclusive.'

Makena Coffman, *Director, UH Mānoa Institute for Sustainability and Resilience, Professor, Department of Urban and Regional Planning, University of Hawaii*

ETHICAL CITIES

*Brendan F.D. Barrett, Ralph Horne
and John Fien*

Routledge
Taylor & Francis Group

LONDON AND NEW YORK

First published 2021
by Routledge
2 Park Square, Milton Park, Abingdon, Oxon OX14 4RN

and by Routledge
52 Vanderbilt Avenue, New York, NY 10017

Routledge is an imprint of the Taylor & Francis Group, an informa business

© 2021 Brendan F.D. Barrett, Ralph Horne and John Fien

The right of Brendan F.D. Barrett, Ralph Horne and John Fien to
be identified as authors of this work has been asserted by them in
accordance with sections 77 and 78 of the Copyright, Designs and
Patents Act 1988.

British Library Cataloguing-in-Publication Data
A catalogue record for this book is available from the British Library

Library of Congress Cataloging-in-Publication Data
A catalog record for this book has been requested

ISBN: 978-0-367-48282-4 (hbk)
ISBN: 978-0-367-48284-8 (pbk)
ISBN: 978-1-003-03904-4 (ebk)

Typeset in Bembo
by Apex CoVantage, LLC

From Brendan F.D.Barrett
to Chizu, Erin and Rian

From Ralph Horne
to Jo

From John Fien
to Esther and Isobel

CONTENTS

List of figures   x

List of tables   xi

Foreword   xii

Acknowledgements   xiv

List of abbreviations   xvii

1   Rationale for ethical cities   1

  What is an ethical city? 1
  How did we get here? 4
  Collective endeavours 6
  Ethical approach to urbanisation 9
  Disconnects between values and cities today 10
  Origins of the ethical city 14
  Revealing the invisible 17
  Outline of the book 18

2   The right *to* the city   24

  Slow emergence of a panoply of rights 24
  Henri Lefebvre and the rise of critical urbanism 26
  Influence of the right to the city movement 29
  The right to the city: where to next? 31
  Rights to the ethical city 34

3 Ethics and the city   41

Searching for ethical urban futures 41
Ethics in contemporary cities 42
Rebuilding on ethical foundations 44
Normative ethics 48
  Consequentialism 49
  Deontological ethics (non-consequentialism) 50
  Virtue ethics 52
Non-normative ethics 53
Is it too late? 56

4 Who shapes the ethical city?   61

Re-configuring cities as ethical 61
Ethical leadership of ideas 62
Ethical urbanists 67
Ethical practitioners in inclusive communities 71
Ethical community entrepreneurs 73
Challenges confronting ethical city shapers 75
Prospects for ethical city shaping 78

5 Assessment of the ethical city   86

Localisation of the Sustainable Development Goals 86
Voluntary local reviews 89
How do we assess progress towards the ethical city? 92
What to measure and how? 97
Need for ethical engagement around the SDGs 100

6 Competitive, liveable and fragile cities   104

Situationism, structure and agency 104
Orientation 1: globally competitive alpha cities 106
Orientation 2: urban liveability 110
Orientation 3: fragile cities 115
Orientation 4: fightback cities – fearless, rebellious and/or
  ethical 119

7 Relentless disruption   124

Utopian and dystopian cities 124
Mega-trends 126
Automation and artificial intelligence 128
Zero-carbon urban futures 132
Disruption and business-unusual solutions 135

8   Building ethical cities   144

    Transformations underway? 144
    New municipalism 145
    Inclusive local economies 150
    Decarbonising local economies 154
    Reimagining cities as ethical 158

9   Transitioning to ethical cities   164

    Ethical realism 164
    Revisiting the key points 168
    Interruptions, disruptions and transitions 169
    Cognizant and responsive ethical city 171
    Actioning transitions to the ethical city 172
    Mediating the means of production 173
    Mediating consumption 174
    Reconfiguring modes of exchange 175
    Processes rather than endpoints 177
    Overcoming fear of the inevitable 178
    Build alliances, share know-how 179
    Conclusion 180

Annex 1: issues covered in the city scan pilot survey   *186*
Index   *192*

FIGURES

1.1	Basic human values derived from Schwartz 2006	12
1.2	Life goals and motivational values	13
3.1	What is the right thing to do?	55
4.1	Arnstein's ladder of participation	73
5.1	Sustainable development goals	87
5.2	VLRs completed or underway in the first cycle of the 2030 Agenda	91
5.3	Cities surveyed in the 2015/2016 Ethical City Scan	95
5.4	Summary of the Barcelona ECS results 2015	96
5.5	Word cloud of additional issues identified by MOOC participants	98
6.1	Mapping connectedness of global cities	107
6.2	Employment structure and functioning of a world city	109
6.3	Number in poverty and poverty rate in the United States: 1959–2018	112
7.1	Average automation potential by US State in 2016	130
7.2	Average automation potential by Australian local government areas in 2030	131
7.3	Deep decarbonisation and negative emissions scenarios compared to emission pledges from the Paris Agreement	134
8.1	Viable emission reduction actions under the Leeds Carbon-Roadmap	155
8.2	Reduction of CO_2 emissions in Berlin 1990–2010 and the scenario until 2050	156
9.1	City of Zurich potential measures to meet 2050 targets	176

TABLES

1.1	Moral values and ethical alignment: Jane Jacobs versus urban planners	15
3.1	Understanding normative and non-normative ethics	45
4.1	Conflicting interests shaping the ethical city	63
4.2	Governing by obeying, Barcelona en Comú Code of Political Ethics	64
4.3	Limits and opportunities for ethical leadership in Barcelona	65
4.4	Ideological approaches and urban planning	68
4.5	Measuring ethics in institutions	78
5.1	Ten principles of the UN Global Compact	93
5.2	Indicator framework for the Ethical City Scan, SDGs and others	93
5.3	Top ten challenges facing cities	97
6.1	Top ten ranked cities in 2019	110
6.2	How cities go feral	116
6.3	Diagnosing ferality	118
7.1	Pressing questions and priority interventions for each scenario	137
8.1	Three types of new municipalism	148
8.2	Changes in levels of general satisfaction with life according to B-MINCOME participation type: 0 = completely unsatisfied, 10 = completely satisfied	153
9.1	Characteristics of ethical realism compared to the Third Way, Old Left and New Right perspectives	167

FOREWORD

It has become commonplace to observe that a majority of the world's population now lives in urban settlements. Cities are the ubiquitous form of social life, the containers within which most people live out their limited time on earth. Of course, ever since the first cities in Mesopotamia, Egypt, the Indus Valley and the mountains of Central and South America the city has punched above its weight. Concentrating political power, economic activity and stimulating innovations, writing, the arts and religious practices, cities have long cast their shadow over humanity and the natural world. Cities are and have always been the source of extreme class differences driving patterns of economic and political inequality. This has been particularly the case since the rise of capitalism, first in the West and now globally.

The authors of this book have chosen to put forward the concept and ideal of 'the ethical city' as a polar opposite to 'the neoliberal city'. The latter has been with us for the past 50 years and is marked by intensified inequality, environmental degradation, lost civility and social exclusion. An ethical city, therefore, looks to a future in which the current urban trajectories of the neoliberal city are taken head-on with the aim of generating radical changes to reverse the slide.

This, they suggest, will require new forms of popular consciousness and mobilisation to challenge, disrupt and subvert business as usual. The emphasis is not on individuals acting ethically but on the incubation of progressive social networks and relationships that collectively transform current urban processes and outcomes. Examples are offered. Readers are invited to come up with their own, such as the appearance of 'sanctuary cities' in parts of the United States to oppose President Trump's anti-immigrant rhetoric and actions.

The first two decades of the twenty-first century have borne witness to an escalating attack on the quality of lives of the vast majority of people in both the Global North and the Global South in the interests of the global one per cent and the elites that carry their bags. The dual public health and economic crisis triggered by Covid-19 has created a need and a space for radical rethinking. This is a brave book whose time has come.

Mike Berry
Centre for Urban Research, RMIT University, Melbourne, Australia

ACKNOWLEDGEMENTS

The research underpinning this book was funded in part from RMIT University's contribution to the UN Global Compact Cities Programme and subsequently by the Osaka University Centre Study of Co* Design (for Brendan F.D. Barrett) including visits to Reading and Totnes in 2019 and to Bristol and Preston in 2020.

We began exploring the notion of the ethical city in 2015. At that time, we launched an ethical city scan survey of signatory cities to the UN Global Compact. Cities that sign up to the principles of the UN Global Compact are required to promote measures in support of human rights, sustainability, worker rights and anti-corruption. We would like to take this opportunity to thank the city officials who kindly responded to the ethical city scan survey. This survey was also supported by an extensive number of colleagues at the Cities Programme including Elizabeth Ryan, Sandra Moye and Michael Nolan.

As a result of the survey, data on 19 cities was collected. Subsequently, we developed a Massive Open Online Course (MOOC) on Ethical Cities including a broad examination of ethical leadership, planning, entrepreneurship and citizenship. As part of the course content, we produced a number of short video vignettes around examples of ethical practice such as the Commons and Nightingale housing projects in Melbourne. We are grateful for the encouragement from Martin Bean, RMIT Vice Chancellor and for the funding for the Ethical Cities MOOC. We also appreciate direct support we received from colleagues at RMIT, Future Learn and from various protagonists/experts in the MOOC production and delivery. This included Andrea Chester, Angeletta Nicoletta, Andrea McLagan, Citt Williams, Crystal Legacy, Jago Dodds, John Postill, Joe Hurley, Katie Vasey, Libby Porter, Lisa De Kleyn, Opie Sayer-Hassall, Wendy Steele, Yolanda Rios, Ross Baxter, Andy Fergus (City of Melbourne), Peter Steele (Hip v. Hype), Meghan Hopper (former Mayor of Moreland Council) and Zac Karvoun (Estate Agent).

This work emerged initially during a period of significant developments at the global level around the New Urban Agenda and the Sustainable Development Goals. We sought to feed the outcomes of our research into the preparatory process for the UN Habitat III Conference. To achieve this, we organised an Urban Thinkers Campus at RMIT University in February 2016, as a contribution to the World Urban Campaign: an advocacy platform facilitated by UN Habitat and bringing together an extensive range of institutions. The Urban Thinkers Campus (UTC) on Ethical Cities – Locking in Liveability took place on 16 February 2016 and addressed three core themes: ethical urban development, urban resilience and inclusion/right to the city. The UTC was well attended with 266 participants from 138 organisations, and the aim was to stimulate debate on how it may be possible to foster shared values for all city dwellers so as to contribute to just and ethical cities. We would like to thank World Vision International for co-hosting the Urban Thinkers Campus, in particular Tim Costello, Joyoti Das, Michael Poustie, Annabel Hart and Jacqueline Treiu.

We greatly appreciate the contribution from a diverse group of speakers and facilitators at the UTC including Steve Chadwick (Mayor of Rotorua), Douglas Ragan (UN Habitat), Martin Thomas (Habitat for Humanity), John Watson (The Conversation), Cezar Busatto (City of Porto Alegre), Robyn Waters (International Real Estate Federation), Austin Ley (City of Melbourne), Toby Kent and Maree Grenfell (Resilient Melbourne), Carmel Guerra Oam (Centre for Multicultural Youth), Ramesh Jumar (AMES Australia), John Van Kooy (Brotherhood of St. Laurence), Danielle Curry (ANZ Bank) and Liz Johnstone (AECOM).

Next, an Ethical Cities–Urban Innovation Forum was organised in Barcelona on 6 July 2016, bringing together 70 European Union, national and local stakeholders for open discussions on key urban challenges and the upscaling of innovative and practical solutions. We appreciate the work of Sergio Tirado-Herrero and his colleagues at RMIT Barcelona in making this event happen.

One of our goals was to ensure that the outcome document from the World Urban Campaign, *The City We Need 2.0*, made specific reference to the ethics and the city. We were successful in contributing to principle one as follows: 'The City We Need is socially inclusive and engaging. The City We Need is people-centred, *ethical*, and just' – our emphasis. This document complements the New Urban Agenda, as negotiated by country representatives, agreed at UN Habitat III, hosted by the city of Quito in October 2016.

To date, through interactions with policymakers, researchers, local leaders, community group leaders and students, we have noted a ready willingness among people to engage with the ethical city. We have observed how many progressive cities are already promoting ethically informed approaches to the challenges they face and how this is tied to positive change on the ground. The aim of this book is to provide a framework in which to embed existing activities within cities so that it may be possible to consolidate efforts around the use of ethics as a guide for local action. We have many colleagues to thank for their guidance and advice including Anitra Nelson, Mike Berry, Colin Fudge, Esther Charlesworth, Tony Dalton and Gavin Wood.

We have had many interesting conversations with a diverse group of people in preparing this book and we would like to thank a few of them here: Gill Ringland, Gurpritpal Singh, Jim Bignal and Emma Borg at Ethical Reading; Genny Gellatly, Jay Tompt and John Elford at Transition Town Totnes; Diana Finch, Managing Director of the Bristol Pound; and Matthew Brown, Preston City Council Leader and Senior Fellow for Community Wealth Building with the Democracy Collaborative.

At Routledge, we would like to thank Andrew Mould for having commissioned this book and Egle Zigaite for her editorial assistance with its production. We also thank Louise Soloway Chan and the Hong Kong Mass Transit Railway Corp for giving permission to reproduce artwork from the Sai Ying Pun metro station on the book cover.

The timing of this book is highly significant. The 2020s is likely to be a tumultuous decade – commencing with the Coronavirus (Covid-19) pandemic, the resolution of Brexit, the US Presidential elections and characterised by ongoing trade wars, the persistence of economic stasis or destabilisation, alongside even more hazardous climatic events. A pessimist may conclude that there would be few bright spots, if any. Seeking a return to the pre-Covid-19 'normal' appears both unviable and unwise. Rather, we need to set a new forward-looking, progressive agenda designed to help those most in need. A book of this nature, raising questions around collective agency and doing the right thing, may be exactly what borderline but nevertheless hopeful pessimists and eternal optimists want to read.

ABBREVIATIONS

AI	artificial intelligence
BAU	business as usual
CCTV	closed-circuit television
CEO	chief executive officer
CLES	Centre for Local Economic Strategies
CNN	Cable News Network
COP	Conference of the Parties
CPTED	Crime Prevention Through Environmental Design
ECS	ethical city scan
EIU	Economist Intelligence Unit
ETB	Edelman Trust Barometer
EU	European Union
GaWC	Globalisation and World Cities
GFC	global financial crisis
GVA	gross value added
HLPF	High-Level Political Forum
ICLEI	International Council for Local Environmental Initiatives – Local Governments for Sustainability
IMF	International Monetary Fund
INESC	Institute for Socio-Economic Research
IPCC	Intergovernmental Panel on Climate Change
ISO	International Organization for Standardization
LA21	Local Agenda 21
LSE	London School of Economics
LVR	local voluntary review
MOOC	Massive Open Online Course
NGO	non-governmental organisation

NUA	New Urban Agenda
NVR	national voluntary review
NYC	New York City
OECD	Organisation for Economic Cooperation and Development
SDSN	Sustainable Development Solutions Network
SDG	Sustainable Development Goals
UBI	universal basic income
UCLG	United Cities and Local Governments
UNDESA	United Nations Department of Economic and Social Affairs
UNODC	UN Office for Drugs and Crime
VCAT	Victorian Civil and Administrative Tribunal

1
RATIONALE FOR ETHICAL CITIES

It looks, indeed, as if we are approaching a period of crisis in urban life; and
Invisible Cities is like a dream born out of the heart of the unliveable cities
we know. Nowadays people talk with equal insistence of the destruction of
the natural environment and of the fragility of the large-scale technological
systems (which may cause a sort of chain reaction of breakdowns, paralyzing
entire metropolises). The crisis of the overgrown city is the other side of the
crisis of the natural world. The image of 'megalopolis' – the unending, undif-
ferentiated city which is steadily covering the surface of the earth – dominates
my book, too.

Excerpt from a lecture delivered by Italo Calvino at Columbia
University in March 1983 (Calvino 2004, pp. 180–181)

What is an ethical city?

When the words 'ethical' and 'city' are bound together, the intention is to promote
collective deliberate ethics at the urban scale, as responses to the major disruptive
forces impacting on cities, and to facilitate organised forms of urbanism that are
ethically governed and conducted.

In the ethical city, collective, community-scale interactions are central in
considering ethical and moral standards. For this book, community is defined in
terms of shared values and reciprocity, rooted in society combining traditional
ethics of justice and rights with feminist and environmental ethics of care, empathy,
responsibility and common good. It requires direct democracy and expression of
public will, amounting to a reinvented demos.

The ethical city implies inclusiveness and universality. It should not perpetuate 'us' and 'them'. This is not the kind of community bound together by
common enemies and a hatred or vilification of others – no amount of shared
inward social support can justify or compensate for such an ultimately antisocial

expression of the idea of community. In the ethical city, community is conditional to inclusiveness.

An ethical city is a place where plans, policies and projects are designed and delivered in such a way as to address core urban concerns in an integrated manner. These core concerns are:

- poverty and inequality;
- governance, democracy and social inclusion and
- sustainability and the climate crisis.

The ethical city concept is not centred upon individuals being ethical in cities. Instead, it is about the relations between people reflected in the common ground between the pursuit of individual rights and collective responsibilities as they address these core concerns.

An ethical approach to urbanism suggests common purpose and solutions. It is not about creating additional layers of bureaucracy to enforce ethical codes and forms of behaviour. Rather it is about connecting these core concerns to aspirations. In turn, this requires a radical repurposing of economic and political processes that facilitate the search for exploration and elaboration of alternative pathways towards a sustainable future and an economy that works for all.

Awareness, transparency and accountability are essential in any ethical city. Those in power are held to account. Corruption and malpractice are addressed through such means, as are social expectations and obligations. Social progress involves finding an appropriate balance between aforementioned individual rights and collective responsibilities. Residents in the ethical city hold each other accountable, so that doing the right thing for and by others and the planet is universally expected and practised. Acting without accountability is antisocial.

The ethical city is not an endpoint; it is an alignment, trajectory or orientation. Definitions of the ethical city should be refined by each community in a dialogue with those in power – local authorities, government agencies, politicians, ruling elites and corporations.

With these characteristics, there can be no blueprint for the ethical city. Every city, region or nation is different and must create its own ethical frame. Nevertheless, ethical cities must exhibit the following core characteristics:

- tackling the three core concerns in an integrated way: (1) poverty and inequality, (2) governance, democracy and social inclusion, and (3) sustainability and the climate crisis;
- respect for human rights, enshrined in independent accessible legal recourse;
- access to decent work, a living wage and a decent place to live and actively opposed to inequality as an antisocial phenomenon; and
- community engagement built upon a foundation of universal respect for human and non-human life and transparent, democratic, accountable and ethical governance.

Ethically oriented cities will ultimately be the ones that succeed in enhancing resilience, improving quality of life, creating productive economies and reducing the environmental burden for all residents.

Our concern is that, in the 2020s and 2030s, cities that fail to work towards these charactcristics will become less attractive, less sustainable, more vulnerable to negative shocks (natural and human-induced) and continuously disrupted by mega-trends over time. They will become dysfunctional and antisocial, as individuals living in them prioritise narrow, short-term interests over those of their community and short-term profits over long-term prosperity (Barrett et al. 2016, p. 12).

In contrast, in the ethical city we anticipate a virtuous cycle, as sustainability is progressively improved, respectful participation by all is valued, encouraged and enabled and a fair go is afforded to all on the basis of rights, merit and effort. People will be drawn to such a city – clean, inclusive, low crime, democratic and rewarding. Following them will be ethical investors, in a high-principled spiral, driven by very different aspirations than the competitive, neoliberal city (Emanuele 2017).

In this book we expand on the ethical city concept and what it offers, providing a rationale for the elements summarised earlier. In our globalised world, where we add close to 1.5 million people every week to the urban population, the notion of a city, town or community as ethical is particularly apt. The word, ethics, derives from the Greek word *ethos* – the accustomed habitat or place to which one returns. It is the root of *ethikos*, is the ability to show moral character and, most fundamentally, ethics which defines the distance from what is and what ought to be. This distance demarcates a space where we have something to do (Certeau 1986, p. 199). Ethics is concerned with what is right, fair, just or good, not necessarily with what is most accepted as normal or expedient (Preston 2014). Ethics is also the source of our hopes, visions, aspirations and dreams. Philosopher Rosalyn Diprose (1991) defines ethics as somewhere that is an action. She argues that to project an ethos is to take a position in relation to others. This is echoed by Walter Jeffko (2018, p. 65), who describes actions as morally right if they promote a personal relation of persons, as opposed to an absolute 'impersonal relation' (without human warmth, feelings or care) of persons. Related to this, the feminist ethical standpoint requires that caring for others be taken seriously (Langlois 2011). The exercise of our human agency to care needs to be embedded in the social structuring of power, priority-setting and decision-making in the city (Bee 1994; Giddens 1984). Beyond human-centric ethics, ideas of environmental ethics direct us also to focus on respect for nature (Nash 1989). Our constant search for a better world is primarily ethical – an ever-expanding circle of human rights protected and injustices addressed. Without ethics there would be no progress since we and our institutions/governance structures would lack a vision to guide us (like a compass), be bereft of principles (our gyroscope) and found wanting of an instrument (our sextant) by which to measure progress. Such tools can help navigate around hedonistic, mean-spirited and corrupt forms of urbanism towards a pragmatism of the possible focused on sophisticated notions of justice, emancipation and collective well-being (Amin 2006).

How did we get here?

Contemporary late capitalism presents a perfect storm of anti-democratic and anti-system politics reflected in popularism, individualism, tribalism and nationalism undermining the viability of traditional institutions and their democratic foundations (Hopkin 2020; Eatwell and Goodwin 2018; Jeffko 2018). For the American political theorist, Wendy Brown, the rise of anti-democratic politics is neoliberalism's monstrous offspring. Neoliberalism, she argues, has dismantled society and dethroned politics while extending the personal, protected sphere with its emphasis on individual freedoms (Brown 2019). It has effectively undermined democracy and created a situation in which political power is exercised by, and for, a part rather than society as a whole (Brown 2015). Moreover, political economist, Jonathan Hopkin, suggests that the rise of anti-system politics on both the right and the left directly opposes the institutions and practices of liberal and social democracy creating space for, and at the same time manipulated by, political entrepreneurs (Hopkin 2020). Five key negative impacts are associated with the rise of neoliberalism: (1) growing inequality, even within wealthy countries; (2) reduced wages and increased debt; (3) redistribution of profits from non-financial companies to the financial sector; (4) growth of personal financial insecurity, destruction of social capital and the resulting rise in crime; and (5) relentless commodification of social and biospheric life/life forms (Benatar et al. 2018, pp. 161–162).

Further, at the same time as *demos* has been disintegrating, communities in a sociological sense are elusive. The philosopher and sociologist, Zygmunt Bauman, described the emergence of communitarian communities in the United States, for example, seeking to counter the negative impacts of overt individualism tied into the neoliberal project. These communitarian communities are essentially an extension of family with a strong emphasis on traditional moral values and functioning as islands of homely and cozy tranquillity in a sea of turbulence and inhospitality (Bauman 2000, p. 182). Elsewhere, new forms of community are emerging, many of which are essentially volatile, transient, single-aspect or single-purpose – a soup kitchen for the homeless, charity clothes giveaway outside the local football ground, community gardening for the lonely, etc. 'Explosive communities' or 'carnival communities', tied into events or spectacles (which appeal to similar interests in otherwise disparate individuals), offer temporary respite from the daily agonies of solitary struggles. The *Anywhere but Westminster* video series by the *Guardian* journalist, John Harris, traces the impacts of austerity and Brexit on people and communities across the UK and frequently remarks on how community groups respond to today's challenges through events because they can no longer rely on politicians and government. These examples of Bauman's explosive/carnival communities break the monotony of our separateness but, sadly, leave everything as is.

The challenge for the ethical city is to encourage city dwellers to rediscover community based on social inclusion, shared values and sense of place. This is easier said than done for two reasons. First, the term community has become increasingly pliable, infinitely malleable, combinable, interchangeable and often devoid of meaning (Poerksen 2004). Second, we could be criticised for employing a rather loose and potentially romantic notion of community, especially in an era when many find ourselves connected to virtual or geographically unbounded communities without ties to the physical places we inhabit. Hence, while we do not advocate a simple and potentially unrealistic model of what constitutes community today, we do think that there may be some characteristics that are universal around how decisions are made, how antisocial behaviour is handled and how people are empowered (Naess 1989).

Societal responses to environmental problems such as climate change often exhibit a range of characteristics that undermine collective action. First of all, there is the issue of Hardin's 'tragedy of the commons' in situations where resources such as the atmosphere are shared but ultimately depleted or polluted as a consequence of collective action or lack thereof (Harding 1968, p. 1245; Wilkenfeld 2016, p. 5). Second, difficulties arise around attaining consensus on the problem and its solution as well as when mediating divergent goals of different community members. Third, major challenges surface, such as breakdowns in trust, when seeking to facilitate action where many are willing to direct and few to act (Smith and Mayer 2018, p. 141). Fourth, in contemporary politics there is often an emphasis on short-term appeasement rather than pursuit of principles based on long-term considerations and also the potential for unethical outcomes achieved by those who shout loudest or best understand political transactions (Jaffe 2018).

Just how people, diverse networks, multiple institutions and changing interests in the ethical city might be reconciled over time is an immensely complicated and ultimately unpredictable task. Nested scales of interaction are at play, from households, neighbourhoods and communities to cities, states and regions (DeMarrais and Earle 2017). Pathways for successful collaboration and power-sharing are shaped by agency and specific interests as well as cultural, historical and environmental factors. It would be futile to attempt to be prescriptive in defining the ethical city or to pursue specific blueprints for action. Rather, we appreciate that understanding local conditions is the essential first step, and this can only be achieved on the ground, city by city.

An outstanding example of an anti-ethical city response was the austerity measures that ensued from the 2008 Global Financial Crisis (GFC), impacting most severely on the United States and Europe (Blyth 2013). Using public money to bail out the banks and corporations reinforced concerns about our increasingly inequitable world, where a few powerful individuals reap the rewards while most people continue to face a constant daily struggle just to get by and pay the bills. The GFC released shockwaves of accelerated economic restructuring, regulatory reorganisation and sociopolitical conflict in cities around the world (Brenner et al. 2012). The same mistake was repeated in the United States, where the Covid-19 stimulus packages and the political, economic, social and human impacts will take years to play out (Goodman 2020; Levitz 2020; Legrain 2020). Consequently, the aftermath of both the GFC and the pandemic reveal the importance of remaining vigilant on how new forms of post-disaster governance take shape and on who controls them (Tooze 2020). At the same time, we have witnessed decades of national inaction on climate change and other pressing sustainability concerns. In the midst of this intractable strife, the ethical city represents a source of inspiration and hope. It provides a way to mend the current divide between people and renew the social contract between city authorities and urbanites. To illustrate how this could happen, we share relevant and actionable ideas on how to positively transform our urban way of life, exploring various teleological end states (maximising good of and happiness for community) and deontological stances (determining right, wrong, just and unjust).

Collective endeavours

Deregulation, marketisation, rising inequality and environmental destruction – in short, the orientation of neoliberalism – promote the idea that people who have resources, conscience and capability are expected to, indeed must, take responsibility for solving the social ills and negative externalities of the free market. These good and able people are alone and individually expected to reduce their carbon footprint, recycle their waste and volunteer their time, intelligence and resources to restore social and environmental justice. According to Peter Bloom, neoliberalism 'individualizes ethics' by making each one of us responsible for dealing with its moral and structural failings (Bloom 2017, p. 103). For example, a Google web search for 'climate change' + 'what you can do' in April 2020 revealed 946 million sites worldwide. Changing 'you' in this search to 'governments' identified 14,300 sites while substituting 'companies' and then 'corporations' for 'you' identified a total of 15,100 sites. These figures are up from 176,000, 594 and 1,334 sites, respectively, from an identical search in April 2008 (Fien et al. 2009). However, despite this predominance and exponential rise, individual agency, choice and actions are limited. This is not to say they are pointless, rather that individualism itself is problematic in that it points away from structural, social explanations of society and pathways to change.

Ethical cities, therefore, are mindful of and take strong action to redress this imbalance between the individual and collective transformative actions required to ensure socially inclusive urban futures. These cities hold the potential for residents to emerge from, or at least, ameliorate, contemporary challenges that many cities are exposed to across the globe. They do not target making residents 'more ethical' in some sort of behavioural change or thought experiment. Such individual-oriented approaches have been extensively tried and tested in the past without much success since they tend to leave unaddressed systemic concerns related to governance and capital. This is not to say that individual action does not matter. As the end titles run in Al Gore's 2006 *An Inconvenient Truth* documentary, we are invited to reduce our carbon emissions to zero, buy energy-efficient appliances, change our thermostats, weatherise our houses, recycle, buy a hybrid car, walk or ride a bicycle and so on. While these are all important actions, as ethical subjects shaped by social structures, such individual actions may perpetuate a capitalist reality whose unsustainable results we may morally oppose or, at least, take issue with (Bloom 2017). On the one hand, we are made responsible for the problem needing behavioural change; on the other, we are expected to accept that we are powerless to ultimately change anything structural. Neither is true.

This conundrum is captured by the term capitalist realism, coined by the British cultural theorist Mark Fisher, describing a situation where political initiatives are only considered realistic when they are capable of operating within the confines of the capitalist system (Fisher 2009). Bursting this false logic requires a shift in the tenets of technologies, rules, knowledge and beliefs that hold unsustainable practices together, including the intellectual and ideological constraints imposed by contemporary capitalism. A myriad of interventions, propositions, actions and alliances is valid in this endeavour, from the smallest act of care to the new social movements fighting in cities against corporate control, hostile immigration policies, gender oppression and ecological devastation (Venturini et al. 2019).

In this context, we envision the ethical city not as *denouement* (the resolution of the story) but as an evolving narrative that shapes our collective urban experience. Hence, an ethical city is not an endpoint but a process and a way of acting. There is no recipe for how, and in what order, realignments and experiments might occur. Moreover, recognising that urban change is unpredictable and contested, it is perhaps inevitable that it is conducted at once by policy innovation within and action from without. Protest movements against social injustices, insensitive responses to pandemics, the accelerating climate crisis and appearance of authoritarian regimes are critical in registering the wider need for change. Their basic premise is that radical, systemic change is required in how and for whom politics and the economy work. A plethora of direct actions at national and local levels promote and catalyse this change (running for election, citizen assemblies, non-violent protests, school strikes, etc.). They form part of a much bigger, global picture of resistance, protest and activism in the face of the profound impacts on urban living resulting from pervasive neoliberalism and unfettered capitalism (Hou and Knierbien 2017). Various slogans sum up the ambition: 'system change, not climate change' and 'cities for people, not for profit', etc. (Brenner et al. 2012).

Needless to say, the ethical city stands in contrast to the neoliberal city, where the former represents a collective utopian aspiration and the latter a form of free market utopianism. The neoliberal city has denied the right to a life of quality and well-being for the vast majority, to support a lifestyle of excess of the tiny minority of elites. In the neoliberal city, local government is modelled on the enterprise, the resident on the consumer and governance on business management (Hackworth 2006). In the ethical city, local government co-creates and co-designs the city with residents, who have both rights and responsibilities, and where governance is a collaborative process. It represents a shift away from neoliberal thinking as elaborated by the economists, Friedrich Hayek and Milton Friedman, which has been criticised for the tacit acceptance of class distinctions and trickle-down economics that has exacerbated wealth inequality (Harvey 2007).

In the ethical city, diverse interests that cannot be truly mediated without conflict are accommodated along the lines of agonistic (rather than antagonistic) urbanism (Mostafavi 2017, p. 13). Inevitably, many of the deepest tensions and contradictions within ethics do not immediately yield easy or straightforward resolution. Offering final answers to every ethical dilemma is not a laudable or useful goal. Rather, the ethical city advances collective thinking on complex problems and in this book it is the scale and frame of the city that is presented as an appropriate vehicle to facilitate necessary exchanges. The politics of proximity within cities requires leaders, businesses, civic groups and communities work hand in hand to resolve their differences. We are already witnessing many urban networks elaborating moral values and ethical principles through initiatives like the fearless city, rebellious city, new municipalism and metropolitan revolutions. To a large degree, we are subscribing here to a form of contextualism through which ethical principles are cognizant of local, regional and national perspectives.

The application of Kantian ethics in local political processes involves the search for a universal pragmatic point of view in the ethical city that is based on the pursuit of workable solutions to contemporary problems. Concurrently, these processes accept the tensions that ultimately emerge when seeking to overlay universalistic ethical frames over more relativistic ethics grounded in contemporary urbanism. The key question is: Under what circumstances might it be possible for the benefits of a universally good and absolute frame of ethics to be sufficiently demonstrated and lived for the diverse, pluralistic communities and sub-cultures within a city to exercise sufficient agency for ethical cities to come into being? Michael Ignatieff's work with gang members in Los Angeles is instructive here. His basic observation is that existing multicultural cities cannot function without a shared moral operating system and moral codes that enable 'millions of people, from different races, origins, and social backgrounds to live together on a daily basis'. Such an operating system, according to Ignatieff, is like the conscience of a city. Moreover, he asserts that 'shared ethics . . . enable people to find common ground when their interests collide' (Ignatieff 2014). How do these shared ethics emerge? Perhaps the best answer is that they have to be negotiated and renegotiated constantly between majorities and minorities, across racial and other divides, across genders and generations, as urban demographics evolve.

Ethical approach to urbanisation

The role of cities in realising human rights through alliances, within cities and across national boundaries, holds considerable promise for global urban justice (Oomen et al. 2016). There have been numerous initiatives across the globe promoting urban sustainability, smart and healthy cities. Our goal here, however, is to document the extent to which ethical urban approaches have already been adopted and to identify some avenues for the future development of these initiatives. We are not directing our observations only at cities in the Global North. While it is true that economic disparities and the marginalisation of large groups of residents are worsening in cities in the Global North, the starker truth is that today slum living is reality for one in every four urban dwellers in the Global South (UN Habitat 2016). This is widely expected to increase to one in three by 2050, with 95 percent of future urban growth predicted to take place in the slums of developing cities in Asia and Africa. Responding to this bleak outlook by addressing inequality geographically and on multiple scales is an important task for any city. What can, and should, be done?

Cities are people. Their collective and heterogeneous fortunes are inevitably the result of the complex interplay of power, shaped by dynamics of technological, economic, political, social and cultural processes. Different cities face many common overarching challenges – variously called mega-trends and wicked problems – including overextended ecological footprints resulting in land, air, water, food and energy insecurity; pandemics, climate change; digital and other disruptive technologies; traffic and clogged urban roadways; post-industrial restructuring; rapid economic change associated with globalisation and marketisation as well as migration and population change. Cities willing to engage residents in positive, integrated, inclusively motivated and sustainable responses to such challenges lie at the heart of how we define progress in the ethical city. For example, new infrastructure is not progress if it exacerbates inequality and/ or climate change by enabling fossil-fuel based consumption and/or restricting access to residents on the basis of their wealth or status.

Given the practical realities that cities face, what ethical frameworks can be used to guide city-shaping and how can these be utilised to enhance urban life for all? What ethical transformations might bring about increased urban resilience, deep decarbonisation and climate change adaptation? Climate responses coupled with appropriate urban governance practices and a renewed focus on addressing inequality and marginalisation in cities are of particular import here, as distinct from climate responses *per se*. Sustainable forms of urban development can only possibly succeed if they incorporate inclusionary form of governance that are capable of responding to the intermix of pressing needs – housing, energy, employment, welfare, water, food, security – of billions of people in both the Global South and North. Even then, there are no guarantees. Reframing the city as ethical also carries an intention to embrace disruption, dynamism and purposive social change. Problematising the unsustainable contemporary city (all of the cities on Earth) carries with it the aim of addressing the many existential threats facing our world rather than running away from them. A search for shared, intrinsic values may make it possible for every city to articulate a moral vision and a set of ethical principles. Logically, if inclusionary governance is to be exercised, these principles must appeal across sociopolitical divides, linking effectively into people's identities and existing world views; in other words, through a true and authentic appreciation of what people consider important to them (Chilton et al. 2012; Lakoff 2004; Marshall 2014).

Disconnects between values and cities today

Today's cities almost universally exhibit endemic and worsening inequality; yet they are also primary incubators of the cultural, social and political innovations that shape our planet (Barber 2013). Cities are simultaneously sites of moral degradation and potential epicentres for ethical innovations. While cities might normatively be viewed as places of enlightened thinking, scientific breakthroughs and technological innovations, they are also hotspots of economic, social and environmental problems. They can be zones of security and safety as well as treacherous places with unsafe streets and crime-ridden districts. Examining empirical evidence, the economist Edward Glaeser found that cross-community mobility in cities makes it difficult to enforce strict ethical urban standards while, conversely, permitting ethical innovations (Glaeser 2000). A greater proportion of city dwellers are more likely to have transgressed against society's more extreme ethical strictures than their rural counterparts. At the same time, city dwellers are also more likely to exhibit progressive ideals and lead in embracing ethical values that are (hopefully) gradually adopted elsewhere. Reiterating the observations made by Jane Jacobs in her 1969 book *The Economy of Cities*, Glaeser concludes the density and the proximity of residents in cities enable them to play an important role in fostering ethical evolutions. This has profound implications for those seeking to promote ethical responses to secure smart, sustainable, resilient and/or liveable cities (Bianchini and Avila 2014; Kitchin 2016).

Although somewhat simplistic in its assumptions, the apparent contradiction between what urban places can exhibit (inequality, unsustainability) and what urban dwellers often think and do (progressive, innovative) opens up questions about relations between urban inequalities, the values of urban dwellers, and the overarching structures that lie beyond their values and weigh heavily on the urban condition. What values do we exhibit? What happens when these values meet society and its structures, and what are the consequences for cities? One agreed-upon conceptualisation of basic values has been formulated by Shalom Schwartz from the Hebrew University of Jerusalem. Schwartz undertook a survey in 70 countries on basic values that individuals in diverse cultures all recognise (Schwartz 2006). This was further refined with additional surveys exploring values along a continuum of compatible and conflicting motivations – personal focus versus social focus, anxiety free versus anxiety avoidance, self-protection versus growth, (Schwartz et al. 2012). The results, subdivided according to ten motivationally distinct value groupings, are shown in Figure 1.1. In this map, values that are most closely related (statistically correlated) to each other are contiguous – hence an individual who attaches importance to conformity is also likely to also consider tradition, security and benevolence to be important. Likewise, an individual who attaches importance on hedonism is likely to thrive on stimulation and achievement.

Another influential survey, undertaken by a team of psychologists led by Frederick Grouzet and Tim Kasser, examined goals (tied into motivational systems) of undergraduate students across 15 cultures. This survey identified 11 goals that can be positioned across two primary dimensions as shown in Figure 1.2 (Grouzet et al. 2005, p. 812). Within this framework, community as a goal can be understood as both self-transcendent (concerned with the welfare of society and future generations) and intrinsic (satisfying the needs for relatedness and affiliation). Applying outcomes from these research surveys, the Common Cause Foundation, an advisory group for charities based in the UK, developed a list of intrinsic values that are relevant to both individuals and collective groups as follows: broadmindedness, a world of beauty, a world at peace, equality, protecting the environment, social justice, helpfulness, forgiveness, honesty, responsibility, self-acceptance, affiliation to friends and family and community feeling (Crompton and Weinstein 2015). Ignatieff identified a similar range of intrinsic values – or what he called ordinary or everyday virtues – from a three-year, eight-nation study in which he interviewed Brazilians, South Africans and Zimbabweans living in informal settlements, farmers in Japan, gang leaders in Los Angeles and monks in Myanmar (Ignatieff 2019). Ignatieff found a common moral language of everyday virtues that could be seen as the moral operating system in major international cities and shantytowns alike: tolerance, forgiveness, trust and resilience. Intrinsic values such as these contrast sharply with extrinsic values, where the emphasis is on wealth creation, social recognition, social status and prestige, control or dominance over people, authority, conformity, preserving public image, popularity, influence and ambition (Crompton and Weinstein 2015). Therefore, a community within an ethical city, which prioritises intrinsic over extrinsic values, is more likely to encourage civic activism around social and sustainability causes.

Such civic activism acknowledges the dire need to promote politically pragmatic, ethical and actionable responses to the challenges facing humanity in this era of planetary urbanisation (Brenner and Schmid 2012). It confronts the scalar limitations of both localism and globalism. Cities are nodes in global networks that may be interrupted, whether by trade wars, pandemics or extreme weather events, but they inevitably involve flows of people, technology, trade, resources and finance, as well as negative externalities such as increasing un- and underemployment and carbon emissions that are often embedded in the production or waste generated from the (over)consumption of goods. Undoubtedly, there has been progress on some of these latter issues in cities around the world. From the

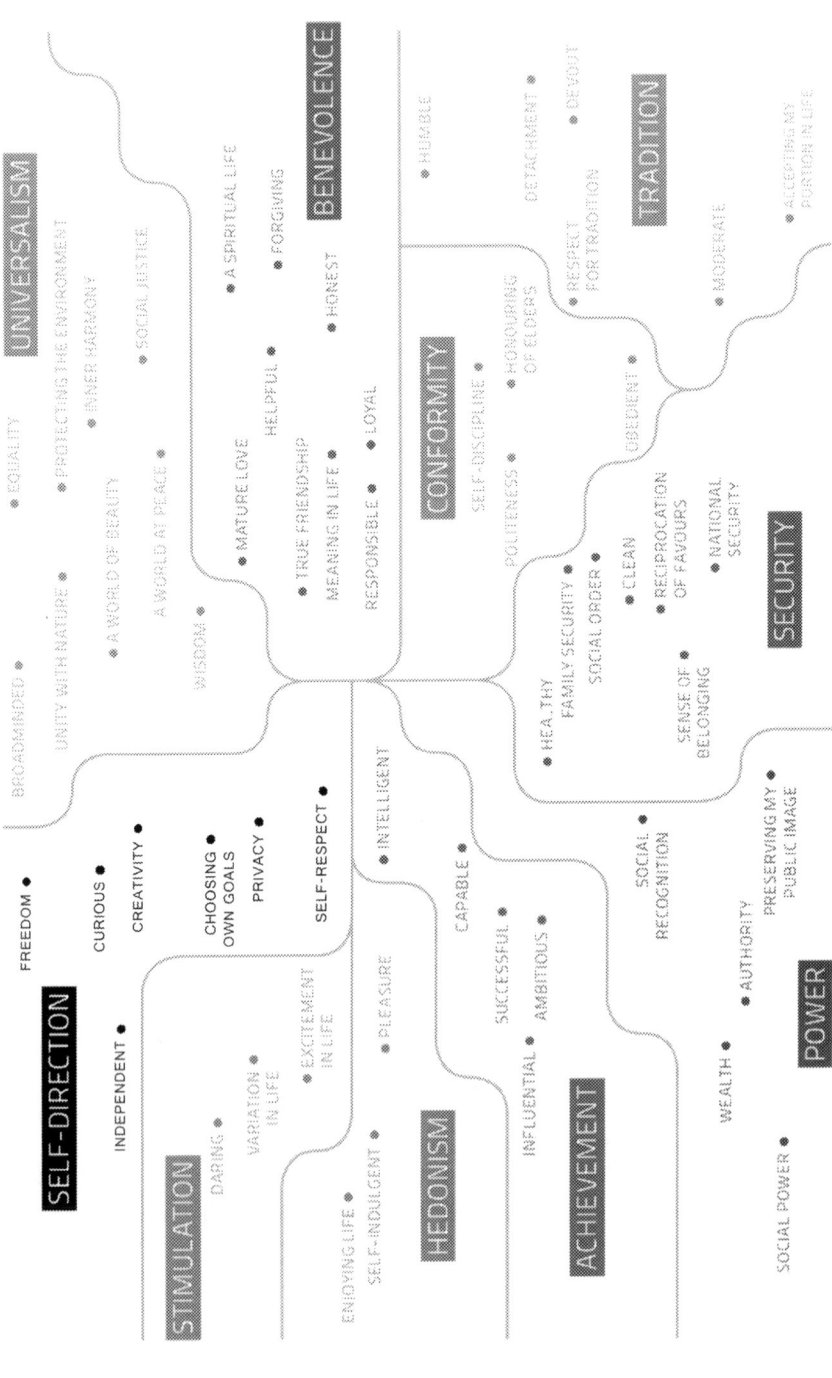

FIGURE 1.1 Basic human values derived from Schwartz 2006

Source: Crompton and Weinstein (2015, p. 122). Design by Minute Works.

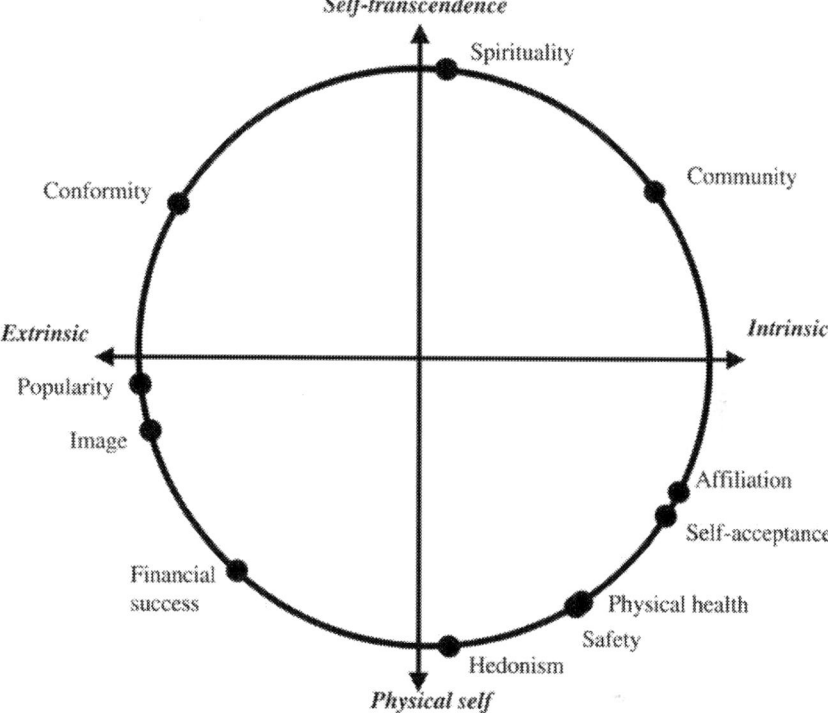

FIGURE 1.2 Life goals and motivational values

Source: Grouzet et al. (2005, p. 812).

1960s onwards, clean air legislation, land use planning and the rise of the welfare state have mitigated the worst excesses of uncontrolled urbanisation and inequality in many cities. Decades of eco-diplomacy have brought intermittent signs of progress on climate change, for example in Paris at COP21 (Conference of the Parties to the UN Framework Convention on Climate Change). However, much of what passes for climate progress has not been integrated with efforts to also tackle inequality and promote good governance, as would be expected in an ethical city.

The disconnects between climate policy and practice by corporate interests on the one hand, and people's views, needs and rights on the other, are played out in cities where much of humanity lives. Cities also function as sanctuaries for the most vulnerable in society (e.g. low income, disenfranchised youth, abused women and girls, refugee migrants, the physically and mentally disabled and the elderly). These groups are most at risk of experiencing urban disasters (whether climate-related or from other risks). Hence, in instances when these vulnerable groups are excluded from urban processes, the relative degree of sustainability and resilience of any given city is put into question. Climate change, pandemics and urban inequality require a 'we are all in this together' position; a 'we first' world, not a 'me first' one (Mainwaring 2011). Measures to deal with inequality and social inclusion are thus prerequisites for all efforts towards sustainable or resilient cities. In this context, tackling inequality and the enhancement of representative governance garner equal billing as essential and interdependent objectives of urban life and planning. These issues (climate, inequality and accountability) have been separated for far too long – a separation that has led to perverse outcomes for city life. For example, rebates to promote solar power are disproportionately accessed by those who are better off even though such rebates are funded by levies on everyone's bills. Moreover, across these domains, failures of successive governments to address climate change and inequality despite repeated promises foster mistrust that is itself a potential conduit for anti-democratic tendencies. In the last decade, austerity has widened inequality further, while big business has recorded historic profits (World Inequality Lab 2018).

Disconnects also abound between the green advocacy of urban dwellers and the physical manifestation of the urban agglomeration as an unsustainable material artefact. Cities continue to consume more resources and produce more pollution on a per capita basis despite decades of policy rhetoric and well-meaning efforts to decarbonise, retrofit and green cities. At the same time, the popularist positing of 'green vs. jobs' dichotomy (or 'job blackmail') (Kazis and Grossman 1991) panders to a disempowered, dispossessed working class in rustbelt, post-industrial regions of the United States and Europe, reeling from decades of post-Freidmanist policy neglect, widening inequality and disrespect (Standing 2011; Hafkamp 2011; Shelby 2016). Inequality cannot be tackled alone; neither can the decline in environmental quality or the growing fragility of democratic and pluralist societies. This requires a means to synchronise competing goals associated with each descriptor. Think of any urban project development process – it has successive design stages. In an ethical city, these would include assessments to ensure that environmental improvements, inclusion and representation aspirations are met in an integrated way. While this might sound simplistic, far-fetched, utopian and naive, the alternatives are bleak in the form of cities that are increasingly undemocratic, unequal and polluted – and facing long-term decline through the draining away of amenity, knowledge and investment.

Origins of the ethical city

Almost every kind of adjective imaginable has been attached to cities in recent years. Examples include mega, global, smart, creative, competitive, multicultural, green, sustainable, low carbon, just, resilient, liveable, happy and healthy. Such titles variously address perceived or real problems and opportunities. Some of the more well-resourced ideas come with detailed criteria defining the overall goal and providing a basis for comparison to determine which is the most resilient city, most sustainable city, most liveable city and so on. In promoting this linear thinking, thus applied, these criteria stigmatise the least resilient, sustainable or liveable (etc.) cities (see Chapter 6). Adjectives matter and have concrete implications. Indeed, a fair portion of the urban lexicon is more or less moulded by or to the interests and concerns of advocacy, policy and corporate institutions, including both established and emerging actors seeking to promote alternatives to business as usual. By advancing the need for integration and trade-offs between priorities for social, economic, environmental and cultural concerns, a significant subset is seeking to counter rampant trends in pollution, consumption and inequality in cities. Given this context and the disruption currently apparent in urbanisation globally, why and where does the ethical city fit? We argue that the ethical city is a term that can do useful work, drawing together interests and repositioning contemporary cross-sectoral priorities for city governance and development. In defining ethical cities, we acknowledge the importance of critical urban theory and development studies (Chapter 2), as well as popular, normative and non-normative ethics (Chapter 3). There is a rich writing tradition on both urban justice and ethical urban practice. In the modern era and in the Global North, ethical dimensions of urbanism are anchored in the pioneering work of Jane Jacobs, who exposed the negative impacts of urban renewal and advanced the concept of social capital in urban sociology (Jacobs 1961). According to Schubert (2019), Jacobs promoted a set of moral values and an ethical alignment that contrasted sharply with her urban planner contemporaries (see Table 1.1).

The origins of the ethical city can be traced back to the archetypical Ancient Greek *polis* – which means both the city and a body of citizens — where, according to Plato, the best form of government would be one that leads to the common good. We acknowledge, however, that this point of origin may not resonate with people from eastern Eurasian, Asian, African, indigenous and other non-Western cultures where ethical values have been shaped by different historical patterns (Lent 2017). That being said, the Singaporean architect William S.W.

TABLE 1.1 Moral values and ethical alignment: Jane Jacobs versus urban planners

Jane Jacobs	Urban Planners
Cities as processes	Cities as a work of art
Organised complexity	Make big plans
Ecosystem	Utopian, modernistic planning
Inductive	Deductive
Concrete, realistic	Abstract
Street-level	Bird's-eye view
Freedom	State control
Self-help	Big government
Bottom-up	Top-down
Pro-urban	Anti-urban
High density	Low density
Urban	Suburban
Conservation, rehabilitation, revaluation	Creative destruction

Source: Schubert (2019, p. 7).

Lim has attempted to develop the notion of Asian Ethical Urbanism, differentiated from Western Modernist planning theory, and placing emphasis on urban projects that enhance happiness, sustainability, justice, rights, equity, wealth and health (Lim 2005). More recently, another Singaporean architect, Jeffrey K.H. Chan, suggested that the ethical city could play an important role in assembling five existing urban discourses into a unified view and creating a new perspective. These discourses are the just city, urban ethics, built environment ethics, ethical dimensions of environmental design and behaviour and ethics in the Anthropocene (i.e. geological epoch marking the commencement of significant human impacts on Earth's geology, ecosystems and climate). For Chan, the fundamental challenge is how to design interventions that envision urban realities, bring about ethically significant consequences and help identify a moral operating system for the city (Chan 2019).

How the physical form of the city shapes what is taken as normal and how this influences urban ethical frameworks has been examined by Canniffe (2005) in the context of urban planning in Britain. Canniffe concluded his research by proposing a methodology for the analysis and design of urban environments with an emphasis on the ethos of shared purpose rather than individual ethics. More recently, Mostafavi (2017) explored relationships between values and the physical aspects of the city (streets, houses, public spaces, etc.) drawing on diverse perspectives from history, art, planning, law and design to elaborate the politics of space. Sennett (2018) elicited ethical insights from contemporary urbanity based on personal practical experience and bringing together various streams of classical thinking and philosophy. He examined the relationship between built environment and the good life through examples drawn from diverse cities, including London, Paris, Mumbai and Medellin. Rather than stooping to allow urbanism to represent society as it is, Sennett questions whether or not in the ethical city we should seek to change urbanism and, as a consequence, change society (pp. 3–4). This has been articulated as the 'right to change ourselves by changing the city' (Harvey 2008, p. 23), which Lefevbre (1996) considered an essential right to the city (as elaborated in Chapter 2). The implication is that modes of urbanisation can be transformed across the globe to maximise the greater good while minimising the negative side effects.

At the time of writing, two European cities had adopted the ethical city as a banner around which to mobilise their community. Ethical Reading, a historic university town in south-east England with a population of 230,000, was conceived in 2017 as a non-profit social enterprise seeking to help businesses in the city become more ethical. In Italy, the city of Prato, located in the Tuscany region with a population of 195,000, adopted the term ethical city with the aim of promoting coordinated approaches to address problems stemming from poverty, lack of housing and reliance on social welfare. While each has adopted the term in different ways to how we present the ideas in this book, these examples do illustrate a broader interest in bringing together ethics and the urban. As a further example, the picturesque market town of Totnes in south-west of England (population of just over 8,000) has been described as one of the most ethical places in Britain (Wilson-Powell 2018). Widely known as the starting point for the Transition Town Movement, this vibrant community experimented with its own local currency and has promoted a range of innovative projects including Caring Town Totnes, Schumacher College, R Economy Centre, Local Entrepreneur Forum and the Totnes Community Development Society. Such initiatives collectively demonstrate attempts to place intrinsic values at the heart of future transformations of the town (Barrett 2019).

The DFG Research Group on Urban Ethics, based at Ludwig-Maximilians-Universität, examined 'good and proper conduct of life in 20th and 21st century cities' with a focus on eight cities – Auckland, Berlin, Bucharest, Istanbul, Moscow, Munich, Singapore and Tokyo – based on data from 2015 to 2018. They identified bottom-up and top-down influences upon urban ethics. The former includes protest movements against social injustice and authoritarian regimes (i.e. Occupy Wall Street) with the latter exercised through various means of governance that include 'appeals to people as ethical subjects and the proclamation of exemplary, good urbanites' (DFG Research Group n.d.). In 2019, an interdisciplinary conference on ethical cities was organised at the D.B. Reinhart Institute for Ethics in Leadership at Viterbo University in the United States, with a focus on how it may be possible to promote ethical living in contemporary urban settings. Ethical living in this context envisaged decision-making shaped by notions of truth, character, consequences and fairness (Kyte 2012), and the conference pamphlet contained a rousing quotation from Aristotle: 'The good of the individual by himself is certainly desirable enough, but that of a nation and of cities is noble and more divine.'

Revealing the invisible

Cities are diverse yet similar; unique yet familiar; obdurate yet always changing. They are not just experienced differently, but they are seen very differently, selectively and partially. As we navigate the city, we often fail to take note of our surroundings. The more we try to focus in and see the less visible, the more it can be obscured. Just as a torch beam focuses our attention on what is revealed along the line of light, it further darkens and obscures all else. With no torch, the darkness and shadows recede. This is not to say that torches do not have their purpose: they both reveal and conceal. So too the tools with which a city is viewed – whether deliberately or inadvertently. The urban high street is seen uniquely by every passer-by. This is not so much consciously as by default; for example, a wholefood vegetarian might note a vegan café as a landmark while people might or might not see the homeless person begging outside a bookstore, supermarket or local café. Noticing spatial polarisation might bring to light class segregation, while physical boundaries might mask this through gated-communities, privatised space or off-limits levels in vertically segregated high-rise buildings (Chapter 6).

For the writer, Italo Calvino, cities can be understood as intersections of ideas and cultural forces, both visible and invisible. What might be obvious to one might be obscure to another. Familiarity also impacts on visibility. Thus, in his classic 1972 novel *Invisible Cities*, Calvino explains how Marco Polo described the 55 cities that he claims to have visited to the great Mongol ruler, Kublai Khan. In his report Polo explicates how:

> the empire is sick, and, what is worse, it is trying to become accustomed to its sores. This is the aim of my exploration: examining the traces of happiness still to be glimpsed, I gauge its short supply. If you want to know how much darkness there is around you, you must sharpen your eyes, peering at the faint lights in the distance.
>
> *(Calvino 1972, p. 51)*

At one point, Kublai becomes suspicious and asks why Polo never talks about his home city, Venice. 'Every time I describe a city, I am saying something about Venice,' Polo replied (ibid., p. 78). Incidentally, Marco Polo's Venice was a Republic ruled by the Doge and a Great Council, composed of hundreds of patricians. A social contract was expressed through visual codes embodied in numerous works of art that adorned the Doge's palace walls tracing the history of the city (Barrett 2015). Venice apparently inspired Jean-Jacque Rousseau's concept of the 'social contract' in 1762, though he considered the Venetian political system to be inferior to that of his native Geneva.

Social contract theory implies that morality is shaped by a set of agreed codes governing how people treat each other for their mutual benefit. Contemporary Venice, marketised, casualised, commoditised and objectified, begs for such a social contract. In comparison, the response to mass tourism in the case of Barcelona offers lessons on how to redress the existing imbalance in favour of tourism by leveraging 'traditions of local urban planning, and [by putting] the rights of residents before those of big business' (Colau 2014). Cities like Amsterdam, Barcelona, Venice, Kyoto and Dubrovnik are also variously recognising the need to renegotiate agreed rules and codes governing how people within their city engage with each other for mutual benefit. This presupposes a new legitimacy in viewing the city as a place primarily for residents rather than a commodity for tourists – with all that this implies for business-as-usual governance and, in turn, for inequality and climate change. The torch beam is shifting, revealing an alternative view of the city.

Outline of the book

In a 1983 lecture at Columbia University, Calvino described his book *Invisible Cities* (1972) as a love poem to the city. In it, he explored the hidden reasons that bring people to live in cities and how urban places enable the exchange of memories, desires and words. He pointed to happy cities that exist in the midst of unhappy cities, to utopian cities (which even if we do not catch sight of, we should not stop looking for) that exist in the midst of infernal cities. In homage, we begin each chapter in this book with a quotation from Calvino.

In Chapter 2 Henri Lefebvre's 1968 book *Le Droit à la ville* provides the starting point for revisiting the many attempts to reclaim the streets for people under late capitalism. From David Harvey to Manuel Castells, the right to the city has had a profound impact upon both urban theory and the manifestations of urban activism since the events in Paris of May 1968. Lessons abound for ethical city activism.

Next, in Chapter 3, the idea of an ethical lens through which cities might be viewed anew is explored with the specific idea of enabling particular new alignments that foster inclusionary solutions to inequality and climate change.

In Chapter 4 the focus shifts more to actors in cities. While the ethical city is not an individual project, and urban leaders cannot, by definition, deliver an ethical city, there are considerations about how capabilities, knowledge and the leadership of such ideas mediate city futures.

In Chapter 5 attention turns to the plans, measurement and monitoring of ethical cities and some of the minefields involved in conventional approaches to planning, doing, acting and evaluating progress. Chapter 6 extends this line of thinking into notions of league tables of cities, exploring four main orientations that frame comparisons within and between cities.

In Chapter 7, the frenetic pace of change in cities is juxtaposed against the incumbencies of privilege and the enduring tools of appropriation. The urban politics of carbon and the disruptions of automation and artificial intelligence provide a backdrop for current struggles to articulate and realise ethical city futures.

Chapter 8 is reserved for examples of grassroots activism in cities that provide fertile experiments in ethical cities, while we conclude the book in Chapter 9 with a discussion on how it may be possible to action the transition towards ethical cities.

References

Amin, A. (2006) The good city, *Urban Studies 43* (5–6): 1009–1023. https://doi.org/10.1080/00420980600676717.

Barber, B.R. (2013) *If Mayors Ruled the World: Dysfunctional Nations, Rising Cities.* New Haven and London: Yale University Press.

Barrett, B.F.D. (2015) Ethical cities are the future, *Our World.* [Online]. Available: http://ourworld.unu.edu/en/ethical-cities-are-the-future [Last accessed: 6 February 2020].

Barrett, B.F.D. (2019) Demise of the Totnes pound won't stop this English town pushing back on austerity, *The Conversation.* [Online]. Available: https://theconversation.com/demise-of-totnes-pound-wont-stop-this-english-town-pushing-back-against-austerity-115484 [Last accessed: 31 January 2020].

Barrett, B.F.D., Horne, R. and Fien, J. (2016) The ethical city: A rationale for an urgent new urban agenda, *Sustainability 8* (1197): 1–14. https://doi.org/10.3390/su8111197.

Bauman, Z. (2000) *Liquid Modernity.* Cambridge: Polity Press.

Bee, H.L. (1994) *Lifespan Development.* New York: Harper Collins College Publishers.

Benatar, S., Upshur, R. and Gill, S. (2018) Understanding the relationship between ethics, neoliberalism and power as a step towards improving the health of people and the planet, *The Anthropocene Review 5* (2): 155–176. https://doi.org/10.1177/2053019618760934.

Bianchini, D. and Avila, I. (2014) Smart cities and their smart decisions: Ethical considerations, *IEEE Technology and Society Magazine 33* (1): 34–40. https://doi.org/10.1109/MTS.2014.2301854.

Bloom, P. (2017) *The Ethics of Neoliberalism: The Business of Making Capitalism Moral.* Routledge Studies in Business Ethics. London and New York: Routledge.

Blyth, M. (2013) *Austerity: The History of a Dangerous Idea.* Oxford: Oxford University Press.

Brenner, N., Marcuse, P. and Mayer, M. (eds.) (2012) *Cities for People Not for Profit: Critical Urban Theory and the Right to the City.* London and New York: Routledge.

Brenner, N. and Schmid, C. (2012) Planetary urbanization, in Gandy, M. (ed.) *Urban Constellations.* Berlin: Jovis, pp. 10–13.

Brown, W. (2015) *Undoing the Demos: Neoliberalism's Stealth Revolution.* New York: Zone Books.

Brown, W. (2019) *In the Ruins of Neoliberalism: The Rise of Antidemocratic Politics in the West.* New York: Columbia University Press.

Calvino, I. (1972) *Invisible Cities.* London: Vintage Books, published in 1997.

Calvino, I. (2004) On invisible cities, *Columbia: A Journal of Literature and Art 40*: 177–182. [Online]. Available: www.jstor.org/stable/41808770 [Last accessed: 7 May 2020].

Canniffe, E. (2005) *Urban Ethic: Design in the Contemporary City.* London and New York: Routledge.

Certeau, M. de (1986) *Heterologies: Discourse on the Other.* Minneapolis and London: University of Minnesota Press.

Chan, J.K.H. (2019) *Urban Ethics in the Anthropocene.* Singapore: Palgrave Macmillan.

Chilton, P., Crompton, T., Kasser, T., Maio, G. and Nolan, A. (2012) Communicating bigger-than-self problems to extrinsically-oriented audiences, *Common Cause Foundation.* [Online]. Available: https://valuesandframes.org/resources/CCF_report_extrinsically_oriented_audiences.pdf [Last accessed: 19 April 2020].

Colau, A. (2014) Mass tourism can kill a city – Just ask Barcelona's residents, *The Guardian.* [Online]. Available: https://www.theguardian.com/commentisfree/2014/sep/02/mass-tourism-kill-city-barcelona [Last accessed: 19 April 2020].

Crompton, T. and Weinstein, N. (2015) Common cause communication: A toolkit for charities, *Common Cause Foundation.* [Online]. Available: Common cause communication: A toolkit for charities [Last accessed: 8 May 2020].

DeMarrais, E. and Earle, T. (2017) Collective action theory and the dynamics of complex societies, *Annual Review of Anthropology 46*: 183–201. https://doi.org/10.1146/annurev-anthro-102116-041409.

DFG Research Group (n.d.) *Urban Ethics*, Ludwig-Maximilians-Universität. [Online]. Available: www.en.urbane-ethiken.uni-muenchen.de/index.html [Last accessed: 6 May 2020].

Diprose, R. (1991) A genethics that makes sense, in Diprose, R. and Ferrell, R. (eds.) *Cartographies: Post-Structuralism and the Mapping of Bodies.* Sydney, Melbourne, Auckland and London: Allen & Unwin.

Eatwell, R. and Goodwin, M. (2018) *National Populism: The Revolt against Liberal Democracy.* London: Pelikan Books and Penguin Random House.

Emanuele, V. (2017) Rebel cities, urban resistance and capitalism: A conversation with David Harvey, *Counter Punch.* [Online]. Available: www.counterpunch.org/2017/02/01/rebel-cities-urban-resistance-and-capitalism-a-conversation-with-david-harvey/ [Last accessed: 23 April 2020].

Fien, J., Moloney, S., Bates, M. and Horne, R. (2009) Carbon neutral communities: A perspective from learning and social change, in Martin, J., Rogers, M. and Winters, C. (eds.) *Climate Change in Regional Australia: Social Learning and Adaptation.* Bendigo: VURRN Press (for Academy of Social Sciences in Australia).

Fisher, M. (2009) *Capitalist Realism: Is There No Alternative?* Winchester and Washington: Zero Books and John Hunt Publishing.

Giddens, A. (1984) *The Constitution of Society: Outline of the Theory of Structuration.* Berkeley: University of California Press.

Glaeser, E.L. (2000) Cities and ethics: An essay for Jane Jacobs, *Journal of Urban Affairs 22* (4): 473–493. https://doi.org/10.1111/0735-2166.00068.

Goodman, P.S. (2020) Why the global recession could last a long time, *The New York Times.* [Online]. Available: www.nytimes.com/2020/04/01/business/economy/coronavirus-recession.html [Last accessed: 1 April 2020].

Grouzet, F.M.E., Kasser, T., Ahuvia, A., Fernandez-Dols, J.M., Kim, Y., Lau, S., Ryan, R.M., Saunders, S., Schmuck, P. and Sheldon, K.M. (2005) The structure of goal contents across fifteen cultures, *Journal of Personality and Social Psychology 89*: 800–816. https://doi.org/10.1037/0022-3514.89.5.800.

Hackworth, J. (2006) *The Neoliberal City: Governance, Ideology, and Development in American Urbanism.* Ithaca: Cornell University Press.

Hafkamp, W. (2011) Assassination in the sustainable city: The Netherlands and beyond, in Charlesworth, E. and Adams, R. (eds.) *The EcoEdge: Urgent Design Challenges in Building Sustainable Cities.* Abingdon: Routledge.

Harding, G. (1968) The tragedy of the commons, *Science 162* (3859): 1243–1248.

Harvey, D. (2007) *A Brief History of Neoliberalism.* Oxford: Oxford University Press.

Harvey, D. (2008) The right to the city, *New Left Review 53* (September–October): 23–53. [Online]. Available: https://newleftreview.org/issues/II53/articles/david-harvey-the-right-to-the-city [Last accessed: 23 April 2020].

Hopkin, J. (2020) *Anti-System Politics: The Crisis of Market Liberalism in Rich Democracies.* Oxford: Oxford University Press.

Hou, J. and Knierbien, S. (eds.) (2017) *City Unsilenced: Urban Resistance and Public Space in the Age of Shrinking Democracy.* London: Routledge.

Ignatieff, M. (2014) The moral operating system of a global city: Los Angeles, *Carnegie Council for Ethics in International Affairs.* [Online]. Available: www.carnegiecouncil.org/publications/articles_papers_reports/0194 [Last accessed: 23 April 2020].

Ignatieff, M. (2019) *The Ordinary Virtues: Moral Order in a Divided World.* Cambridge: Harvard University Press.

Jacobs, J. (1961) *The Death and Life of Great American Cities.* New York: Random House.

Jacobs, J. (1969) *The Economy of Cities.* New York: Random House.

Jaffe, C. (2018) Melting the polarization around climate change politics, *The Georgetown Environmental Law Review 30*: 455–497. [Online]. Available: www.law.georgetown.edu/environmental-law-review/wp-content/uploads/sites/18/2018/07/melting-_GT-GELR180017.pdf [Last accessed: 28 April 2020].

Jeffko, W.G. (2018) *Contemporary Ethical Issues: A Personalist Perspective,* 4th Edition. New York: Humanity Books.

Kazis, R. and Grossman, R. (1991) *Fear at Work: Job Blackmail, Labor, and the Environment,* 2nd Edition. Gabriola Island: New Society Publishers.

Kitchin, R. (2016) The ethics of smart cities and urban science, *Philosophical Transactions A: The Royal Society Publishing 377* (2083). https://doi.org/10.1098/rsta.2016.0115.

Kyte, R. (2012) *An Ethical Life: A Practical Guide to Ethical Reasoning.* Winona: Anselm Academic Publishing.

Lakoff, G. (2004) *Don't Think of an Elephant! Know Your Values and Frame the Debate.* Vermont: Chelsea Green Publishing.

Langlois, L. (2011) *The Anatomy of Ethical Leadership: To Lead Our Organizations in a Conscientious and Authentic Manner.* Edmonton: Athabasca University Press.

Lefebvre, H. (1996 [1967]) The right to the city, in Kofman, E. and Lebas, E. (eds.) *Writings on Cities*. London: Blackwell.

Legrain, P. (2020) The coronavirus is killing globalization as we know it, *Foreign Policy*. [Online]. Available: https://foreignpolicy.com/2020/03/12/coronavirus-killing-globalization-nationalism-protectionism-trump/ [Last accessed: 16 April 2020].

Lent, J. (2017) *The Patterning Instinct: A Cultural History of Humanity's Search for Meaning*. New York: Prometheus Books.

Levitz, E. (2020) What's good, bad and mediocre in the coronavirus stimulus bill, *New York Magazine*. [Online]. Available: https://nymag.com/intelligencer/2020/03/coronavirus-stimulus-bill-explained-bailouts-unemployment-benefits.html [Last accessed: 1 April 2020].

Lim, W.S.W. (2005) *Asian Ethical Urbanism: A Radical Postmodern Perspective*. Singapore: World Scientific Publishing Company.

Mainwaring, S. (2011) *We First: How Brands and Consumers Use of Social Media to Build a Better World*. New York: Palgrave Macmillan.

Marshall, G. (2014) *Don't Even Think about It: Why Our Brains Are Wired to Ignore Climate Change*. New York: Bloomsbury.

Mostafavi, M. (ed.) (2017) *Ethics of the Urban: The City and the Spaces of the Political*. Zurich: Lars Muller Publishers.

Naess, A. (1989) *Ecology, Community and Lifestyle*. Translated and Edited by David Rothenburg. Cambridge: Cambridge University Press.

Nash, J. (1989) *The Rights of Nature: A History of Environmental Ethics*. Madison: University of Wisconsin Press.

Oomen, B., Davis, M.F. and Grigolo, M. (2016) *Global Urban Justice: The Rise of Human Rights Cities*. Cambridge: Cambridge University Press.

Poerksen, U. (2004) *Plastic Words: The Tyranny of a Modular Language*. University Park, PA: Penn State University Press.

Preston, N. (2014) *Understanding Ethics*, 4th Edition. Alexandria: The Federation Press.

Schubert, D. (2019) Jane Jacobs, cities, urban planning, ethics and value systems, *Cities* 91: 4–9. https://doi.org/10.1016/j.cities.2018.05.001.

Schwartz, S.H. (2006) Basic human values: Theory, measurement, and applications, *Revue Francaise de Sociologie* 47: 249–288.

Schwartz, S.H., Cieciuch, J., Vecchione, M., Davidov, E., Fischer, R., Beierlein, C., Ramos, A., Verkasalo, M., Lönnqvist, J.-E., Demirutku, K., Dirilen-Gumus, O. and Konty, M. (2012) Refining the theory of individual basic values, *Journal of Personality and Social Psychology* 103 (4): 663–688. https://doi.org/10.1037/a0029393.

Sennett, R. (2018) *Ethics for the City: Building and Dwelling*. New York: Allen Lane, an imprint of Penguin Books.

Shelby, T. (2016) *Dark Ghettos: Injustice, Dissent and Reform*. Boston: Belknap Press.

Smith, E.K. and Mayer, A. (2018) A social trap for the climate? Collective action, trust and climate change risk perception in 35 countries, *Global Environmental Change 49*: 140–153. https://doi.org/10.1016/j.gloenvcha.2018.02.014.

Standing, G. (2011) *The Precariat: The New Dangerous Class*. New York: Bloomsbury Academic.

Tooze, A. (2020) The normal economy is never coming back, *Foreign Policy*. [Online]. Available: https://foreignpolicy.com/2020/04/09/unemployment-coronavirus-pandemic-normal-economy-is-never-coming-back/ [Last accessed: 15 April 2020].

UN Habitat (2016) *World Cities Report 2016*. Nairobi: UN Habitat.

Venturini, F., Degirmenci, E. and Morales, I. (eds.) (2019) *Social Ecology and the Right to the City*. Quebec: Black Rose Books.

Wilkenfeld, J. (2016) *Myth and Reality in International Politics: Meeting Global Challenges through Collective Action*. London: Routledge.

Wilson-Powell, G. (2018) This is why Totnes one of the most ethical places in Britain, *Pebble Magazine*. [Online]. Available: https://pebblemag.com/magazine/travelling/totnes-independent-ethical-shops-guide [Last accessed: 22 April 2020].

World Inequality Lab (2018) World inequality report 2018, World Inequality Database. [Online]. Available: https://wir2018.wid.world [Last accessed: 23 April 2020].

2

THE RIGHT *TO* THE CITY

> In my *Invisible Cities*, Kubla Khan is a melancholy ruler who realizes that his boundless power is of little worth because the world is rapidly going downhill. Marco Polo is a visionary traveller who tells Khan tales of impossible cities.
>
> Excerpt from a lecture delivered by Italo Calvino at Columbia
> University in March 1983 (Calvino 2004, p. 179)

Slow emergence of a panoply of rights

Rights to free speech, to assembly and to protest are basic human rights; yet they are routinely removed, undermined, minimised and dispersed by forces of power. They are valuable, necessary and therefore not benignly granted. Property rights, voting rights and patriarchal accommodations all raise ethical questions about rights in the city and the sources of power that determine, maintain and/or undermine them. Even in the rich western nations, post-Industrial Revolution rights have been hard won and remain contingent. The enfranchising of millions of working men in Britain occurred just over 150 years ago with the 1867 Reform Act. Women over 30 in Britain attained suffrage with the 1918 Representation of the People Act, but universal suffrage had to wait another decade. Post-World War II rebuilding saw struggles for rights *in* the welfare state, such as the rights to life, livelihood and an adequate standard of living, health and housing. This manifested internationally with the 1948 Universal Declaration of Human Rights (Howie 2018). These rights, of course, fell unevenly, leaving women, children, indigenous people and a mass of so-called minorities to struggle for rights to freedom from discrimination on the basis of gender, age, race, disability, sexual orientation and beyond.

Fast forward to the last half century and a veritable mountain of critical urban scholarship has accumulated since Henri Lefebvre published *Le Droit à la ville* in 1968 (Lefebvre 1996). Harvey, Castells, Marcuse and others have led the way in spatial interpretations of social studies on capitalism. Critical urban theory thus enables analysis of ideology, of power, inequality, injustice and exploitation within and among cities (Marcuse 2011; Brenner 2009). Within this, as David Harvey's seminal work, *Social Justice and the City* shows, the right to the city relates to the ability of residents to use urban space and to be represented in and through it (Harvey 1973). For example, he questioned the priority given in many cities to the rights of private vehicle owners and the impact this has on shaping urban space. The perverse incentives that favour investments in roads and expressways, even though they exacerbate inequality and climate change, mean that most urban residents still remain locked into the internal combustion engine even though it is clearly an outdated technology. As Harvey illustrates, an emphasis on equality, democracy and diversity would require that car ownership be understood as just one of many interests of urbanites, rather than the pre-eminent one. However, this remains elusive, as existing dominant rights *in* the city continue to usurp attempts to secure rights for all *to* the city. Put simply, these existing rights are granted through capitalist processes – elites who reproduce and entrench the primacy of power to hold land and property and to exploit other human resources to create profit and accumulate more capital. Thus configured, rights *in* the city are currently distributed in uneven ways that undermine universal rights *to* the city from being effectively established. Moreover, as Lefebvre warned, the ruling elite will always try to suppress contestation of those rights and will always subvert romantic possibility (our dreams of a city that works for all) into the disappointments of an all-too-realistic actuality (Merrifield 2006).

Reductions or unevenness in rights have extensive effects and affects upon residents and other users of urban space, facilities and services (Marcuse 2010). In the neoliberal capitalist city, property rights take precedence and even determine how liberal-democratic rights can be accessed. A homeless person or a squatter (or anyone else for that matter) lacks the right to occupy unused or empty housing and the right to legal status through an address. Laws and practices of public participation in planning and development proposals favour wealthy landowners over residents (Thorpe 2017). Furthermore, 'property rights of owners outweigh the use rights of inhabitants, and the exchange value of property determines how it is used much more so than its use value' (Purcell 2014, p. 142). The resultant lack of rights *to* the city amounts to the curtailment of the right:

> to change ourselves by changing the city. It is, moreover, a common rather than an individual right since this transformation inevitably depends upon the exercise of a collective power to reshape the processes of urbanization. The freedom to make and remake our cities and ourselves is . . . one of the most precious yet most neglected of our human rights.
>
> *(Harvey 2008, p. 23)*

Reclaiming the urban is as urgent as ever. In a world where surveillance is cheap and ubiquitous, personal data can be easily traded, and the infringement of rights can extend well beyond the city boundary (Cardullo et al. 2019). Informational rights to the city are contested, with Google Maps as a good example (Shaw and Graham 2017). The power of Google and other mega-tech companies is well established as lacking basic accountability. Furthermore, post-truth inequality and discrimination remain deeply ingrained yet agile, as rights dwindle with regard to access to a whole gamut of urban services and participation in the instruments of city governance. Long-standing struggles for rights to a clean, unpolluted environment are joined by those for urban low-carbon mobility, energy and water. In this chapter, while interrogating the right to the city and linking with the idea of the ethical city, we maintain the semantic, but very real, distinction between rights *to* and *in* the city. We also explore the ways in which rights *to* the city as a movement have been shaped in cities today, for better or worse. Then we move onto examine what we can learn from the right *to* the city as a set of purposive ideas and how these insights can in turn inform the ethical city.

Henri Lefebvre and the rise of critical urbanism

Henri Lefebvre's 'the right *to* the city' was proposed as 'a cry and a demand' for people to determine their own destinies through transformed, renewed citizenship and participation in urban life. It represents the 'right to live out the city as one's own, to live for the city, to be happy there. . . . The right to have your urban horizon as wide or as narrow as you want' (Merrifield 2017). While Lefebvre's ideas influenced many urban theorists from the 1970s onwards, the right *to* the city did not gain traction in the English-speaking world until the 1990s with translations of Lefebvre's later works such as *The Production of Space* (1991) and *Writings on Cities* (1996). In these, the central sociological critique of cities as both reflections and co-creators of the societies within which they find themselves takes centre stage. Ideas that are now long established in critical urban studies were built upon the observations that cities are shaped by conflicts between individuals and communities of interest, governments and private business. Lefebvre saw his home city of Paris mirroring the capitalist and class relations contained within it – and manifesting them in segregated spatial patterns of wealth and poverty and the displacement and marginalisation of the poor and other excluded groups (Tomley and Hobbs 2015, p. 106).

Working at the same time in Detroit and Toronto, Mario Bunge led 'expeditions' to research the impacts of the geography of everyday life and drew similarities to Paris. In his *Fitzgerald: The Geography of a Revolution* (1971), Bunge exposed the spatial patterns of rat bites, broken glass, empty blocks, crime and car accidents in cities and their impact on the health and life opportunities of poor, particularly African American, children. He found tenements and high-rise public housing towers that provided none of the amenities and conviviality of Le Corbusier's *Cité radieuse* in Marseilles. Children had no space in which to play and had to improvise, meaning their public spaces were often abandoned lots, busy streets and dangerous alleys. These conditions persisted and later, Mitchell (2003) argues, 'the right to inhabit the city thus demands more than just houses and apartments: it demands the redevelopment of the city in a manner responsive to the needs, desires, and pleasures of its inhabitants' (p. 21). Bunge's teams also investigated the effects of the changing nature of cities, including the expansion of freeways manifesting the rights of machines – cars – over people (echoing Harvey) and gentrification of the inner city, reflecting the privileging of 'foreign invaders' in the form of outside capital, incoming gentrifying residents, etc. over local residents whose rights to fully inhabit the city were being revoked by market forces.

Over the decades from the 1960s to the present, the forces of capital usurped racial discrimination and other pre-existing structures by which privilege is asserted. This is not to say that racism, patriarchy and other far-right leanings disappeared, but that the structures favouring capital power in the city swung decisively in its favour over this period. With it, widespread alienation arose while access to genuine (free) public spaces fell; streets turned to toll roads; footpaths provided and controlled by developers; parks and recreation areas taken over by businesses and many seemingly public places, such as shopping centres, arcades, galleries and the forecourts and foyers of office and residential towers, actually being subject to the rules and, increasingly, the security cameras and uniformed guards of proprietors. Alongside this shift, publicly owned places have been lost to private development, and where retained, often turned over to corporate sponsored festivals and activities (Minton 2009). More pervasive, Martin (2019) argues such trends are contributing to placelessness, with communities increasingly psychologically lost and unable to recognise themselves through the 'conspicuous theming, a lack of local identity or by branding with naming rights' of both public and private spaces in cities.

David Harvey (2008) famously lamented such developments and illustrated through myriad examples how the right to the city was not just 'falling into the hands of private or quasi-private interests' but being *given* to them. He cited actions led by Mayor Bloomberg to reshape New York along lines favourable to transnational interests, financiers, developers and tourists. Similarly, re-cobbled downtown streets in Mexico City suit the 'tourist gaze' and universities have contributed to gentrification by redeveloping inner urban areas – think Yale in New Haven, Johns Hopkins in East Baltimore, Columbia in New York and the Savannah School of Design in Savannah (Bose 2015). Despite neighbourhood opposition in all these cases, local people found their rights to the city restricted in order to advance 'the interests of a small political and economic elite who are in a position to shape cities more and more after their own desires' (Harvey 2008, p. 38).

Others have equated the right to the city with the transcendent ideal of justice reflected in the right to participate in the creation of the city (Fainstein 2014; Purcell 2008; Marcuse 2009). In elaborating on the importance of the just city, Susan Fainstein argues that urban governance systems can – at least in principle – shape environments that enable social inclusion, justice and care relative to others through the pursuit of deliberative development strategies (Fainstein 2010). Analysing New York, London and Amsterdam, Fainstein argued that equality, democracy and diversity are three criteria that aid the assessment of a city as just. Such ideas also inform Julian Agyeman's exploration of the potential to promote just sustainability in urban policy, planning and practice (Agyeman 2013). According to Agyeman, conditions for just sustainability include (1) improvements in quality of life and well-being; (2) intra-generational and intergenerational equity; (3) justice and equity in processes, procedures and outcomes; and (4) living within ecosystem limits. Applying notions of justice in an urban context, Hutson focuses on the struggle of ethnic minority, low-income communities for economic, social and environmental justice in the North American cities of Boston, Brooklyn, San Francisco and Washington (Hutson 2015). Hutson elaborates the strategies and actions employed by these communities in the face of pressures for urban renewal and gentrification from city economic development agencies, real estate developments and other powerful interests. His key conclusion is that we need place-based community development strategies, programmes and policies in order to address growing inequality and injustice in our cities.

There is a long history of campaigns for urban environmental justice that extends back to clean air and water movements in the Global North and to the rapid rise in global environmental awareness sparked by researcher-activists such as Rachel Carson in *Silent Spring* (Carson 1962). Research and advocacy for environmental justice was initiated in the United States in Robert D. Bullard's *Dumping in Dixie* (Bullard 1990). In this regard, it is not surprising that the organisation which developed out of the Love Canal tragedy is called the Centre for Health, Environment and Justice (Mah 2013). The reality of so many neighbourhoods of the poor and racial minorities co-located with toxic waste sites or freeway overpasses is a fundamental issue in urban ethics and raises serious questions around the right to the city.

The right to the city has a fundamental basis in the analysis of inequality, in particular, based upon property. This aligns closely with gender politics, and feminist urbanism has become an essential movement exposing how gender bias and violence against girls and women present widespread obstacles to the right *to* the city. Urban planning has been described as a distinctly masculinist domain with an urban planning gender gap. Women are disproportionally represented patrons of urban public transport, and the impacts of privatised, declining so-called 'public' mobility systems fall heavily along gender lines, undermining women's rights *to* the city, whether for work or recreation. This deeply affects the way women, 'who run most of the world's households – experience the city' (Sassen 2016). In addition, women disproportionately experience sexual harassment and abuse on the streets and in the workplace and work double-shifts of poorly paid jobs and unpaid domestic work. As Falú and Sassen (2017) lament, the violence against women 'that takes place in cities goes beyond robbery and assault, the gang that controls the corner, the abuses, the drug ring that terrorises the neighbourhood . . . [and] the illegitimate use of force'. Almost 100,000 women were intentionally killed in 2017, more than half of them by intimate partners or family members (UNODC 2018). This raises the question of whose right *to* the city we are talking about if such conditions are allowed to persist.

Influence of the right to the city movement

Lefebvre's concept of the right to the city took shape through and influenced his interactions with a group of Marxist architects, sociologists, philosophers and urbanists known as Utopie who were active in the period from 1967 to 1978. This group challenged the rationale underpinning post-war modernisation and urban planning (Buckley and Violeau 2011). Also influenced by Lefebvre, another group of urban theorists, including Castells, Harvey, Marcuse and others, were inspired by the goal of injecting a spatial element into Marxist theory. Castell's early writings provide a Marx-informed analysis of urbanism examining the impact of capital accumulation and other factors on the urban condition across the globe (Castells 1977). Departing from the orthodox Marxist focus on social inequality within the mode of production, he sought to also present the city as a site of social reproduction. He viewed social movements as important agents for the sort of reforms that would meet the needs of residents and involve them more extensively in local decision-making (Castells 1983). His ideas developed to incorporate networks, digital communications and collective consumption patterns that reshape politics, economy, culture, social relations and city functions (ibid.). Critical urban theorists and urban geographers now span a broad field from the more theoretically eclectic to activists committed to Lefebvre's right to the city concept (Fainstein 2014, p. 3).

Consequently, even in the twenty-first century, in the face of the ongoing assault of capital on the city, calls for the right to the city remain undiminished. For example, in 2005, Habitat International Coalition proposed the World Charter on the Right to the City, advocating rights to co-existence; to live with dignity in the city; to habitat that facilitates a network of social relations; to social cohesion and the collective construction of the city; and to influence municipal government (Habitat International Coalition 2005). In turn, this spawned campaigns on housing, transport, public spaces, health and so on and renewed claims for representation and participation in planning and development. Related initiatives include the Right to the City Alliance, the European Charter for Human Rights in the City, the Global Charter-Agenda for Human Rights in the City, and the Gwangju Guiding Principles for a Human Rights City. A veritable plethora of such initiatives inevitably faces critique. So many right-to-the-city initiatives amount to splintered 'calls' – albeit loaded with the moral obligations of what 'cities should do', mixed with rage and righteousness – that in the main have fallen on the deaf ears of capital and administrative bodies. Note also the ambiguous agency in the word 'cities' in these calls, which generally fail to disentangle people, material, power and the machinery of governance – not to mention the many hierarchies of departments and responsibilities in city councils. Those lacking such fulsome consideration inevitably also overlook the limits on power (and resources) of local councils alongside that of regional, state or national governments and their corporate allies. What is left after such calls recede, except the reification of Lefebvre's principles? Indeed, without new ideas as to the course of the interventions in the city that might prompt structural shifts that tip up forces of capital or at least cause elites to pause, any attempts to institutionalise the 'right to the city' moniker are ironic, even oxymoronic (Lamarca 2009).

Most recently the right to the city became a central issue in the negotiations leading up to the agreement of the New Urban Agenda (NUA) at the UN Habitat III conference in October 2016. This document, agreed by national governments, was meant to be a visionary plan designed to influence urban development over the next 20 years. The preparatory process for this UN agreement was complemented by extensive civil society engagement that resulted in a parallel document entitled *The City We Need 2.0.* While civil society organisations, and some local and national governments, pushed for the NUA to place the right to the city at the centre (De Paula 2016), all that they succeeded in gaining was an acknowledgement that:

> We share a vision of cities for all, referring to the equal use and enjoyment of cities and human settlements, seeking to promote inclusivity and ensure that all inhabitants, of present and future generations, without discrimination of any kind, are able to inhabit and produce just, safe, healthy, accessible, affordable, resilient and sustainable cities and human settlements to foster prosperity and quality of life for all. We note the efforts of some national and local governments to enshrine this vision, referred to as right to the city, in their legislation, political declarations and charters.
>
> *(United Nations 2017)*

This led to concerns that the NUA effectively narrowed the right to the city concept towards the realisation of human rights in cities (Turok and Scheba 2018). A number of commentators argued against including the right to the city in the NUA. Purcell (2014, p. 142) argued that contemporary initiatives aimed at entrenching the right to the city in a statist framework misunderstand the right to the city as merely a 'proposed addition to the list of existing liberal–democratic rights'. On the other hand, Coggin (2018) argued that incorporation in the NUA would have a number of positive implications for the right to the city including an important advocacy function, a leverage function (whereby sub-national entities are able to facilitate their claims to the right to the city) and the potential for this non-binding international agreement to be translated into domestic policy in some countries. With such a major opportunity lost, where does that leave the value of the right *to* the city in the ethical city?

The right *to* the city: where to next?

Peter Marcuse (2013) suggests we may have been too literal in interpreting Lefebvre's ideas and that we might usefully take a more pragmatic view, or a strategic one, or one of, variously: a discontented reading, a spatial reading, a collaborationist reading, a subversive reading. Indeed, since all have strengths and weakness, why not adopt multiple such lenses to achieve an integration of the alternative interpretations? Such an integration includes a critique of the primarily economic determinism we saw in the aetiology of the problems of the mid-twentieth-century Paris, Toronto and Detroit and which continue to exist today in, for example, the privatisation of urban public spaces. However, it also draws attention to Lefebvre's focus on everyday life in urban spaces, as a way to counter economistic ideology and thus theorizes people as more than class actors, political sites beyond the workplace and historical forces in addition to economic production (Purcell 2014, p. 145).

Lefebvre's focus on the lived spaces that people actually experience opens opportunities for a wide range of interventions. Some can be led by the state and/or city councils, others by grassroots groups (see, for example, Purcell's (2016) notion of 'democratic publics without the State'). Some can be overtly political, indeed revolutionary; others may focus on cultural rights, promoting tolerance and environmental care while others might involve neighbourhood collectives for sharing childminding, tools and garden allotments. They can even extol hacking smart city infrastructure as a 'bottom-up opening up of top-down government structures and procurement processes' as an exercise of rights *to* the city (de Lange and de Waal 2019, p. 3). The important thing in such acts of city remaking is that, through them, we reshape ourselves as individuals and communities, and we do this in ways that have both a spatial and a political meaning. Thus, Marcuse (2013) proposes a three-step process, which builds on desired existing patterns and processes that serve the wider public good but seeks to change power relations that determine how they will be used:

- **Expose**: Analysing the roots of the problem and making clear and communicating that analysis to those that need it and can use it;
- **Propose**: Working within community to develop and implement actual proposals for strategies that address root causes of problems;
- **Politicise**: Communicating the political implications of what was exposed and proposed through both broadcast media and new digital platforms.

As Marcuse states:

> Moving beyond the use of the right to the city in theory, in charters, and as slogans, it is vital to look at how it has been used *in practice* by organisations directly calling for its implementation as their purpose.
>
> *(Marcuse 2010a, p. 4; emphasis added)*

For example, the Expose–Propose–Politicise strategy can be seen in the way community-based organisations in many cities are promoting actions by women to reclaim-the-night, including the Girl Gangs Over programme in Germany which uses feminist street art against street harassment. Photographic images of girls and women armed and staring at the street with an aggressive expression are placed on walls, windows etc. so that women can reclaim agency in the face of the malestream construction of the city as a place of fear.

The same three-fold strategy can be seen in Egypt where over 95 percent of women report having been harassed sexually in public places, and Cairo is rated the world's most dangerous city for women (Kanso 2017). With social taboos keeping women and men from talking about this problem, crowdsourcing has been used to counter gender-based harassment and violence in Cairo. Harass-Map uses web and mobile technologies to crowdsource incidents with survivors or witnesses reporting occurrences through a web interface, Facebook or Twitter. The reports are automatically mapped by location and colour (according to the nature of the event) and made publicly available on the HarassMap website (see https://harassmap.org/en/). The locations, dates and times of reported incidents appear together with a text description of the incident. These range from brief notes to several pages. These descriptions and viewer responses to them can be seen when the incident dot on the map is clicked. Each report also receives a response with information on how to access free legal services and psychological counselling. HarassMap is not only a key tool for generating data and warnings about dangerous places in Cairo but also a political platform for challenging ste-reotypes and misinformation about sexual harassment in Egypt (Lundahl 2014).

Of course, women's rights *to* the city include many issues beyond safety. For example, women have been instrumental in resisting (so-called) urban renewal projects in many cities and in the struggle for housing rights. Development and housing policies in Turkey since the early 2000s provide a case in point. Gülçin Erdi (2018) has provided a deeply anthropological account of women's resistance to the Ankara Metropolitan Municipality and its plans to 'upgrade' the Dik-men Valley area of the city. Much of this area comprises informal settlements or *gecekondu* (literally, built overnight). Long-term residents were to be moved to a peripheral area that not only lacked infrastructure, services and employment, but it would also require displaced *gecekondu* residents to take out expensive loans for relocation and access to new property. Although ethnic, religious and political differences had divided and dispersed the residents, thus making mobilisation difficult, meetings were held, street representatives elected and a neighbourhood assembly formed of these representatives. The assembly led the mobilisation and negotiation process through an 'Office for Housing Rights' in opposition to the Municipality's project office, which they called the 'Demolition Office'.

Sit-ins in front of the national parliament were organised, media statements issued in front of the municipality of Ankara, and rallies and meetings were held in the central square of Ankara along with an annual festival and numerous lawsuits against the Municipality. The leadership and widespread participation of women were significant features of these actions. One activist, Nazli, is quoted as saying, 'I don't want to leave my house. We, women, were expected to be subordinate, silent, and obedient even against injustice. . . . Now it is over, we want our right to shelter after all these painful years' (Erdi 2018, p. 103). Another, Sultan, stressed the importance of neighbourhood: 'A house outside of my neighbourhood means nothing to me. The neighbourhood is as important as my house. We are all poor, oppressed. The neighbourhood is the place that unites us all' (ibid., p. 108). As a result of this multifaceted and sustained resistance, the Dikmen Valley neighbourhood remains thus far principally intact.

These examples are illustrative of the great many 'modest interventions' (Sassen et al. 2018) that show us how the right *to* the city is being (re-) claimed by marginalised groups, in the case of these examples, women and girls, in cities around the world. Many hundreds, perhaps thousands, of stories of modest interventions exist and span a wide range of actions by community-based organisations seeking to reclaim and assert their rights *to* the city. These include actions related to, for example, the right, especially of the homeless, to use open space as a site for living, the right to housing, squatters' rights, the right to employment, rights to health, rights to a clean environment, rights to information, rights to meaningful participation and so on. Numerous case studies showcase these many stories of social mobilisation for urban rights, justice and a clean environment. They also analyse the motivations of protagonists, the strategies they used, the sources of the opposition and challenges they faced, the compromises they may have had to make and the eventual outcomes and long-term impacts of their actions (Sugranyes and Mathivet 2010; Domaradzka 2018; Yip et al. 2019).

It would never be our intention to criticise the aspirations and actions of any community activism to claim the right *to* the city. Indeed, all of us, the authors, have been involved in campaigns related to urban rights, ranging from the right to protest, to climate justice and campaigns for local organic food over many decades. However, we have been very conscious that any successes we attained were always limited to the specific areas of focus of individual campaigns. We have felt like we were, to use an unfortunate militaristic metaphor, 'winning battles' but never 'winning the war' because it is very hard to address the root cause of the problem: the dominance of neoliberal policies in the capitalist city. Peter Marcuse (2013) suggests that a preferable strategy might be for urban activist groups to combine their efforts and work for urban rights sector by sector, for example, housing and then women's safety, and then food security, and then employment, then migrant rights and so on. This follows his ideas on building socialism one step at a time (Marcuse 2010b). However, advocates of what has come to be known as the 'New Municipalism' have developed a different strategy as discussed in Chapter 8 (Russell 2019; CLES 2019). This involves actors in the wide range of urban social movements working together in tactical networking across the diverse fields of social interest and activism and then collaborating 'to elect progressive candidates to municipal office' and, eventually, gain control of local councils (Thompson 2020). This reflects Harvey's view that exerting the right to the city 'inevitably depends upon the exercise of a collective power over the processes of urbanization' (2013, p. 4) but within the context of a merging of the vertical power of councils with the horizontal power of local community organisations (CLES 2019, p. 7).

Rights to the ethical city

The right to the city has currency as a means to expose, propose and politicise. Lefevre's focus on urban space brings attention to the questions of specificity: whose rights, with what consequence, what rights and in which city and which neighbourhood? Such qualities are essential to the ethical city as a platform for reordering or reshaping the city around responses to climate change and inequality, built on democratic forms of governance and accountability. This raises the question of collective agency of the ethical city. How might ethical city ideas enhance the work of the right to the city and alter the course of urban history?

To begin with, this requires that we recognise how critical urban theory needs to be open to new perspectives (Jayne and Ward 2017). For example, Fujita (2013) suggests that urban theory lacks a crisis perspective and that we need new theories that help us understand the impact of crises, their aftermath and their impacts on cities. While she was specifically referring to the 2008 Global Financial Crisis (GFC) and makes reference to climate change, her assessment is very relevant to cities post-pandemic. In addition, to this we would add concerns around the disruptive impacts of technology (as examined in Chapter 7). It is not that critical urban theorists are unaware of these challenges. Harvey and Wachsmuth (2011) address the irrationality of capitalism in times of crisis (referring to the GFC aftermath) and ask 'What is to be done? And who the hell is going to do it?' The mechanics of just how political mobilisation on a mass scale will occur are highly contested. Again and again, actions, however successful locally, however joined and however politicised, face the brute force of pro-elite structures and the incumbency of the forces of capital. For the right to the city movement it has been thus since 1968:

> The events of 1968 are mentioned recurrently . . . as manifesting, simultaneously, the transformative potential and the endemic difficulty of united, collective action across diverse constituencies. The possibility for such action is further constrained by the potent force of the corporate media, the daily, routinized language of politics, and the perceived need to deal with everyday crises before long-term, systemic issues can be addressed. And, above all, transformative action is constrained by the propaganda of market fundamentalism, the induced appeal of mass consumerism, the technically instrumentalized educational system, the oppressive weight of bureaucracy, and through it all, the overwhelming force of dominant ideologies of exclusion and supremacy.
>
> *(Brenner et al. 2012, p. 7)*

The point we make here is that while there will always be constraints imposed by the incumbency of the forces of capital, there will also always be new unpredictable circumstances that create a gap here and a gaping hole there in pro-elite structures. It is by taking advantage of these opportunities when everything is in flux that it may be possible to bring about positive transformations. Contemporary collective action is diverse, ranging from single-issue platforms to wider class-based resistance, to more overt urban orientations. The climate crisis, for example, poses both risks and opportunities. In the climate changing city where, if you are poor and female and from a 'minority community', then, needless to say, you are far more likely to be a victim of a heat wave than those who can afford to manoeuvre themselves to the safety of air conditioning. In this way, the climate crisis exacerbates already existing inequalities in the city and raises ethical questions about interspatial inequities between the wealthy and the urban poor. The victims are often innocent bystanders who, already poor, find themselves in the path of the worst extremes of climate change. Ethical concerns thus articulated (Gardiner and Hartzell-Nichols 2012) follow familiar lines from previous work outlining how uneven residential segregation in cities (Harvey 1973) results in the various social classes in cities being 'unequally able to protect their environmental interests' (Logan and Molotch 1987, p. 58). In many cities heatwaves have already become the biggest killer of all environment-related disasters (Tuccillo and Buttenfield 2016). The great heatwave of 2018 in Europe caused 30,000 deaths, up to five times more likely than in the past (Carrington 2019). The heat island effect exacerbates the very urban epicentres where inequality is greatest and where rights to the city are most eroded, i.e. where the most vulnerable live (Radford 2017).

In resisting the climate crisis and its uneven effects, front-line solidarity is just as legitimate and essential as linking such actions across cities and to political platforms of wider resistance. A political sensibility attributes climate change induced excessive urban heat to structures of production and consumption. Moreover, it also attributes the uneven effects on morbidity and mortality to the same structures that put particular residents in harm's way. The fact that disadvantaged and low-income households not only suffer immeasurably more but also generate much lower greenhouse footprints in the first place is perverse. Equally perverse is the fact that dominant sanctioned economic levers for reducing energy consumption are targeted at consumption pricing, which has a disproportionately negative effect upon disadvantaged and low-income households, forcing them through brute poverty out of using whatever means of heating and cooling they do have and leaving those who are wealthy and therefore not price-sensitive unaffected (McGann and Moss 2010).

These perverse urban realities, like those described by Lefebvre, Bunge, Harvey and others, reflect normative ethics, which we will discuss further in Chapter 3, as being concerned with 'what is the right thing to do in a given situation.' As such, they are a rebuke to the nihilism of capitalism-led urban development and its negative impacts on the urban fabric and the well-being of all residents. The right *to* the city ought to be 'the right to transform the city into something radically different' (Harvey 2009). The challenge is to identify the ethically right thing/s to do to redress this situation. Lefebvre's conceptual framework for this challenge posits the city as an *oeuvre*, a work in progress, in which all its people can participate; '[t]he right to the *oeuvre*, to participation and appropriation (clearly distinct from the right to property), are implied in the right to the city' (Lefebvre 1996, p. 174). Separating property rights *in* the city from use value rights *to* the city, Lefebvre advances participatory democracy and the removal of obstructions and impediments to collective action.

Acknowledging the huge contribution of the right to the city movement and the centrality of the need to attend to pro-elite structures and the incumbency of the forces of capital, the ethical city is about building visibility into doing the right thing in the face of and with cognisance of these structures and forces with both transparency and accountability. Urban futures are dynamic yet obdurate, fluid yet structured, and crises of urban reproduction present points of ethical intervention. In many cases, and most worrisome, such crises are taken as opportunities by forces of control to reframe rights to the city against the interests of the residents, rolling back initiatives designed to address climate change, inequality or to enhance accountability and transparency. Whether it be pandemics, extreme weather events or economic crises, such moments are times to be vigilant: both to act to protect rights, ethics and the ground gained and to advance to enhance protections and liberties of residents.

Much research and reportage on rights *to* and justice *in* the city tends to focus on the efficacy of policy and action in terms of who benefits and who loses. This differs significantly from an ethical city focus on how each place and each community goes about articulating a moral vision, a set of ethical principles and defined values that form the basis of effective responses to the major issues of today. The ethical city can be utilised to strengthen and bolster the right to the city in three ways: First, by making more visible and overt the idea, and by demonstrating and manifesting ethical practice in the city (and beyond). Second, emergent crises such as climate change and inequality/ poverty are core issues to fight for, but they must also be fought for together; otherwise, we risk unintended consequences and the injustice of addressing climate change but further exacerbating poverty, etc. It is possible to suggest that the right to the city may promote a dis-aggregation of rights and fragmentation of responses to such issues. The ethical city could promote more integrated responses by focusing on what is the right thing to do and who is responsible. Finally, at the heart of the ethical city is the question of how to ensure transparency and accountable governance. Potentially, the attainment of both may be much more possible, but paradoxically harder, as we enter the social media and post-truth era. This requires that ethics are placed firmly at the core of everything we do, especially in the face of ongoing and relentless disruption as we will discuss in Chapter 7.

References

Agyeman, J. (2013) *Introducing Just Sustainabilities: Policy, Planning, and Practice*. New York: Zed Books.

Bose, S. (2015) Universities and the redevelopment politics of the neoliberal city, *Urban Studies 52* (14): 2616–2632.

Brenner, N. (2009) What is critical urban theory, *City 13* (2–13): 198–207.

Brenner, N., Marcuse, P. and Mayer, M. (eds.) (2012) *Cities for People Not for Profit: Critical Urban Theory and the Right to the City*. London and New York: Routledge.

Buckley, C. and Violeau, J. (2011) Utopie: Texts and projects, 1967–1978, in *Semiotexte*. Cambridge: MIT Press.

Bullard, R.D. (1990) *Dumping in Dixie: Race, Class, and Environmental Quality*. Boston: Westview.

Bunge, W. (1971) *Fitzgerald: Geography of a Revolution*. Athens and London: The University of Georgia Press.

Calvino, I. (2004) On invisible cities, *Columbia: A Journal of Literature and Art 40*: 177–182. [Online]. Available: www.jstor.org/stable/41808770 [Last accessed: 7 May 2020].

Cardullo, P., Di Feliciantonio, C. and Kitchin, R. (2019) *The Right to the Smart City*. Bingley: Emerald Publishing Limited.

Carrington, D. (2019) Climate change made European heatwave at least five times likelier, *The Guardian*. [Online]. Available: www.theguardian.com/science/2019/jul/02/climate-change-european-heatwave-likelier [Last accessed: 10 May 2020].

Carson, R. (1962) *Silent Spring*. Boston: Houghton-Mifflin.

Castells, M. (1977) *The Urban Question: A Marxist Approach*. London: Edward Arnold.

Castells, M. (1983) *The City and the Grass Roots*. Berkeley: University of California Press.

Castells, M. (1988) *The Informational City: Information Technology, Economic Restructuring and the Urban-Regional Process*. Oxford: Blackwell Publishing.

Centre for Local Economic Strategies (2019) *New Municipalism in London*. Manchester: CLES. [Online]. Available: https://cles.org.uk/wp-content/uploads/2019/04/New-Municipalism-in-London_April-2019.pdf [Last accessed: 11 May 2020].

Coggin, T. (2018) Law & the New Urban Agenda: A role for the right to the city, *Fordham Urban Law Journal*. [Online]. Available: https://news.law.fordham.edu/fulj/2018/05/11/thomas-coggin-law-the-new-urban-agenda-a-role-for-the-right-to-the-city/#_ftn81 [Last accessed: 30 March 2020].

de Lange, M. and de Waal, M. (eds.) (2019) *The Hackable City: Digital Media and Collaborative City-Making in the Network Society*. Dordrecht: Springer.

De Paula, N. (2016) The right to the city and the New Urban Agenda, *SDG Knowledge Hub*. [Online]. Available: http://sdg.iisd.org/commentary/policy-briefs/the-right-to-the-city-and-the-new-urban-agenda/ [Last accessed: 30 March 2020].

Domaradzka, A. (2018) Urban social movements and the right to the city: An introduction to the special issue on urban mobilization, *VOLUNTAS: International Journal of Voluntary and Nonprofit Organizations 29*: 607–620.

Erdi, G. (2018) Woman and resistance in urban space, Alternatif Politika, *Cities: Identities, Appropriation of Space and Resistance Practices*. [Online]. Available: https://halshs.archives-ouvertes.fr/halshs-02165194/document [Last accessed: 10 May 2020].

Fainstein, S.S. (2010) *The Just City*. Ithaca: Cornell University Press.

Fainstein, S.S. (2014) The just city, *International Journal of Urban Sciences 18* (1): 1–18. https://doi.org/10.1080/12265934.2013.834643.

Falú, A. and Sassen, S. (2017, updated 2018) Women and the city: Reclaiming the streets to impose equal rights, *The Conversation*. [Online]. Available: https://theconversation.

com/women-and-the-city-reclaiming-the-streets-to-impose-equal-rights-88279 [Last accessed: 29 April 2020].

Fujita, K. (2013) *Cities and Crisis: New Critical Urban Theory*. Sage Studies in International Sociology. London: Sage Publications.

Gardiner, S.M. and Hartzell-Nichols, L. (2012) Ethics and global climate change, *Nature Education Knowledge 3* (10): 5. [Online]. Available: www.nature.com/scitable/knowledge/library/ethics-and-global-climate-change-84226631/ [Last accessed: 10 May 2020].

Habitat International Coalition (2005) World charter on the right to the city. [Online]. Available: http://hic-gs.org/document.php?pid=2422 [Last accessed: 29 April 2020].

Harvey, D. (1973) *Social Justice and the City*. Athens and London: Johns Hopkins University Press.

Harvey, D. (2008) The right to the city, *New Left Review 53*: 23–53.

Harvey, D. (2009) David Harvey at the world social forum, Belem. [Online]. Available: www.hic-gs.org/document.php?pid=2953 [Last accessed: 10 May 2020].

Harvey, D. (2013) *Rebel Cities: From the Right to the City to the Urban Revolution*. London: Verso Books.

Harvey, D. and Wachsmuth, D. (2011) What is to be done? And who the hell is going to do it? In Brenner, N., Marcuse, P. and Mayer, M. (eds.) *Cities for People, Not for Profit: Critical Urban Theory and the Right to the City*. London and New York: Routledge.

Howie, E. (2018) Protecting the human right to freedom of expression in international law, *International Journal of Speech-Language Pathology 20* (1): 12–15. https://doi.org/10.1080/17549507.2018.1392612.

Hutson, M.A. (2015) *The Urban Struggle for Economic, Environmental and Social Justice*. Abingdon and New York: Routledge.

Jayne, M. and Ward, K. (eds.) (2017) *Urban Theory: New Critical Perspectives*. Abingdon and New York: Routledge.

Kanso, H. (2017) *Cairo Named Riskiest Megacity for Women, Worse since Arab Spring*, Thomson Reuters Foundation. [Online]. Available: https://poll2017.trust.org/stories/item/?id=535e9698-7e22-4b5f-805e-9077c3d08e43 [Last accessed: 10 May 2020].

Lamarca, M.G. (2009) The right to the city: Reflections on theory and practice, *Polis*. [Online]. Available: www.thepolisblog.org/2009/11/right-to-city-reflections-on-theory-and.html [Last accessed: 10 May 2020].

Lefebvre, H. (1991) *The Production of Space*. Translated by Donald Nicholson-Smith. Oxford: Basil Blackwell, originally published in 1974.

Lefebvre, H. (1996 [1968]) The right to the city, in Kofman, E. and Lebas, E. (eds.) *Writings on Cities*. London: Basil Blackwell.

Logan, J. and Molotch, H. (1987) *Urban Fortunes: The Political Economy of Place*. Berkeley: University of California Press.

Lundahl, E. (2014) Cairo women put sexual harassment on the map, *Humanosphere 9*. [Online]. Available: www.humanosphere.org/human-rights/2014/10/cairo-women-put-their-assaults-on-map/ [Last accessed: 10 May 2020].

Mah, A. (2013) Lessons from Love Canal: Toxic expertise and environmental justice, *Open Democracy*. [Online]. Available: www.opendemocracy.net/en/lessons-from-love-canal-toxic-expertise-and-environmental-justice/ [Last accessed: 10 May 2020].

Marcuse, P. (2009) Postscript, in Marcuse, P., Connolly, J., Novy, J., Olivo, I., Potter, C. and Steil, J. (eds.) *Searching for the Just City*. New York: Routledge.

Marcuse, P. (2010a) Rights in cities and the right to the city? in Sugranyes, A. and Mathivet, C. (eds.) *Cities for All: Proposals and Experiences towards the Right to the City*. Santiago de Chile: Habitat International Coalition.

Marcuse, P. (2010b) In defence of theory, *City, 14* (1–2): 4–12. https://doi.org/10.1080/13604810903529126.

Marcuse, P. (2011) From critical urban theory to the right to the city: What right, whose right, to what city, how? in Brenner, N., Marcuse, P. and Mayer, M. (eds.) *Cities for People, Not for Profit: Critical Urban Theory and the Right to the City.* London: Routledge.

Marcuse, P. (2013) *Reading the Right to the City*, Blog #40. [Online]. Available: https://pmarcuse.wordpress.com/2013/11/20/blog-40-reading-the-right-to-the-city/ [Last accessed: 10 May 2020].

Martin, C. (2019) Wrong division: The rise of privately-owned public space, *Foreground.* [Online]. Available: www.foreground.com.au/public-domain/wrong-division-the-rise-of-privately-owned-public-space/ [Last accessed: 10 May 2020].

McGann, M. and Moss, J. (2010) *Smart Meters, Smart Justice? Energy, Poverty and the Smart Meter Rollout, Social Justice Initiative.* Melbourne: University of Melbourne.

Merrifield, A. (2006) *Henri Lefebvre: A Critical Introduction.* New York and London: Routledge.

Merrifield, A. (2017) Fifty years on: The right to the city, in A Verso Report (ed.) *The Right to the City.* London: Verso Books.

Minton, A. (2009) *Ground Control: Fear and Happiness in the Twenty-First-Century City.* London: Penguin.

Mitchell, D. (2003) *The Right to the City: Social Justice and the Fight for Public Space.* New York: The Guilford Press.

Purcell, M. (2008) *Recapturing Democracy.* New York: Routledge.

Purcell, M. (2014) Possible worlds: Henri Lefebvre and the right to the city, *Journal of Urban Affairs 36* (1): 141–154.

Purcell, M. (2016) For democracy: Planning and publics without the state, *Planning Theory 15* (4): 386–401. https://doi.org/10.1177/1473095215620827.

Radford, T. (2017) Urban heat islands lave cities sweltering, *Climate News Network.* [Online]. Available: https://climatenewsnetwork.net/urban-heat-island-leaves-cities-sweltering/ [Last accessed: 7 May 2020].

Russell, B. (2019) Fearless cities municipalism: Experiments in autogestion, *Open Democracy.* [Online]. Available: www.opendemocracy.net/en/can-europe-make-it/fearless-cities-municipalism-experiments-in-autogestion/ [Last accessed: 17 March 2020].

Sassen, S. (2016) Built gendering, *Harvard Design Magazine 41.* [Online]. Available: www.harvarddesignmagazine.org/issues/41/built-gendering [Last accessed: 10 May 2020].

Sassen, S., Pieterse, E., Bhan, G., Hirsh, M., Falú, A., Ichikawa, H., Riffo, L., Tan, P. and Tarchopulos, D. (2018) Cities and social progress, in IPSP (ed.) *Rethinking Society for the 21st Century: Report of the International Panel on Social Progress.* Cambridge: Cambridge University Press.

Shaw, J. and Graham, M. (2017) An informational right to the city, in A Verso Report (ed.) *The Right to the City.* London: Verso Books.

Thompson, M. (2020) What's so new about New Municipalism? *Progress in Human Geography*: 1–16. https://doi.org/10.1177/0309132520909480.

Sugranyes, A. and Mathivet, C. (eds.) (2010) *Cities for All: Proposals and Experiences towards the Right to the City.* San Diego: Habitat International Coalition.

Thorpe, A. (2017) Between rights in the city and the right to the city: Heritage, character and public participation in urban planning, in Durbach, A. and Lixinski, L. (eds.) *Heritage, Culture and Rights: Challenging Legal Discourse.* Oxford: Hart Publishing.

Tomley, S. and Hobbs, M. (eds.) (2015) *The Sociology Book.* London: Dorling Kindersley.

Tuccillo, J. and Buttenfield, B. (2016) Model-based clustering of social vulnerability to urban extreme heat events, *Geographic Information Science 8*: 114–129.

Turok, I. and Scheba, A. (2018) 'Right to the city' and the New Urban Agenda: Learning from the right to housing, *Territory, Politics and Governance 7* (4): 494–510. https://doi.org/10.1080/21622671.2018.1499549.

UN (2017) *New Urban Agenda*, Habitat III Secretariat. [Online]. Available: http://habitat3.org/wp-content/uploads/NUA-English.pdf [Last accessed: 10 May 2020].

UNODC (2018) *Global Study on Homicide 2018: Gender-Related Killing of Women and Girls.* [Online]. Available: www.unodc.org/documents/data-and-analysis/GSH2018/GSH18_Gender-related_killing_of_women_and_girls.pdf [Last accessed: 10 May 2020].

Yip, M., López, M. and Sun, X. (eds.) (2019) *Contested Cities and Urban Activism*. Basingstoke: Palgrave Macmillan.

3

ETHICS AND THE CITY

> Kublai reflected on the invisible order that sustains cities, on the rules that decreed how they rise, take shape and prosper, adapting themselves to the seasons, and then how they sadden and fall into ruins.
>
> Italo Calvino, *Invisible Cities*, 1972, p. 110

Searching for ethical urban futures

Ethical quandaries emerge from every urban design decision – from prioritising car traffic over pedestrians (and vice versa); roads over rail; single use shopping malls over mixed use development; urban surveillance over respecting privacy and the many other choices that city politicians and planners have to make. The ever-evolving reality of cities is that they are places where problems have to be resolved and where, inevitably, new ones routinely appear. Disruptive smart city technologies, for example, have ethical implications associated with self-driving cars, automation of work and infiltration of artificial intelligence into every sector. To intervene for 'better outcomes' involves an exploration of reciprocities between urban design and its social, cultural and political consequences (Mostafavi 2017, p. 16). The application of ethical principles in intensely politicised urban contexts involves a search for socially inclusive futures and the potential for cities to address the current state of precarity that many residents are exposed to across the globe (Standing 2011; Das and Randeria 2015).

Cities will continue to confront an array of challenges as they decarbonise, manage demographic change and try to cope with evolving patterns of economic activity and technological change. They will grapple with severe negative impacts associated with digital platforms (cyberbullying, online shaming, trolling), e-democracy (election hacking) and human rights (battles over the right to own our data). Historic trends suggest that some of today's most successful cities will face declining economic vitality and deteriorating liveability. At the same time, other cities will grow too quickly for their infrastructure and public services to keep pace. Some will advance to become global cities while others will be characterised by increasing inequality, poverty, homelessness and squalor.

Regardless of whether growing or declining, a large number of cities could, if they continue on their current trajectory, become fragile, fragmented, feral and failing (Muggah 2014a, 2014b, 2016). We already see this manifesting in many cities today with increased inequality, violence, social exclusion, unemployment and disenfranchisement from political processes. This has arguably exacerbated narrow populism, new political movements and reality-television politicians feeding off a fear that Wendy Brown describes as 'no future for the white men' and which is characterised by feelings of nihilism, fatalism and resentment on the part of those who are losing status and wealth (Brown 2019, pp. 161–188). The urban crisis, however, is global in nature and has the potential to generate extreme political responses. Addressing this fear and its consequences entails the pursuit of ethical approaches to urban life, economics and governance. An ethic of unbridled neoliberalism has brought us to this point, although it has generally been operating below the radar and invisible to most people. This is a sloppy form of ethics primarily concerned with maintaining the status quo of power relations rather than securing ethical futures.

Ethics in contemporary cities

Ethics lies at the heart of every political decision made in our cities. Our proposition for the ethical city should not be interpreted as inferring that people, government officials and urban institutions are somehow lacking in ethics. Rather, we maintain that the ethical city stands in contrast to the neoliberal city. This is not to insinuate that the latter is devoid of an ethical frame. The starting point is to recognise that neoliberal ethical reasoning is present in almost every city and community today, and that this is highly problematic, as we will explain later. Moreover, ethics in the contemporary city is contested by diverse interest groups with divergent demands. There is a tradition of ethics in the city involving 'a set of principles and rules of conduct that ensure the safeguarding not so much as the built environment in itself as the people living in it and their well-being' (Moroni 2013, p. 198). These principles can be broken down into institutional ethics (concerning the action of local governments and other institutions) and professional ethics (concerning the actions of land-use planners, architects, policymakers, etc). The basic aim of institutional ethics is harm avoidance, maximisation of the benefits of governmental actions, stewardship of resources, ensuring fairness and serving justice, honouring commitments and maintaining legitimacy (Welch 2014). Professional ethics, on the other hand, focus on a duty to correctly apply knowledge, avoid deceit, avoid conflicts of interest, refuse bribes and other benefits designed to bias decisions, be impartial and to seek the highest quality outcomes from decisions

(Thompson and Leidlein 2008). Almost every city today has a code of ethics (e.g. the Los Angeles code was first promulgated in 1959), ethical guidelines, anti-corruption procedures and an ethics commission with responsibility to address irregularities in election campaign financing, governmental ethics, lobbying practices and contracts. As the Los Angeles *Ethics Handbook for City Official* points out:

> [E]thics laws were adopted to help preserve the public trust by promoting elections and government decisions that are fair, transparent, and account-able. To further those goals, City officials are held to appropriately high standards of conduct, to help ensure that they act in the best interests of the public and the City.
>
> *(Los Angeles City 2019, p. 4)*

While these are all very important considerations, they do not present an overall ethical orientation for the city as a whole, involving all urban stakeholders working collaboratively.

There have been numerous research initiatives evaluating the effectiveness of ethical practice within local governmental institutions. After observing hours of official meetings, conducting interviews, executing surveys and organising focus groups, a five-year study of local government in Denver, USA revealed that the emergence of an ethical culture within an organisation is influenced by the degree of communication around ethical concerns and the extent to which this dialogue involves all stakeholders (Jovanovic and Wood 2006). Similarly, research in Australia discovered that when there is limited discourse on ethical concerns within the planning profession it results in a noticeable lack of ethical leadership (Cook and Sarkissian 2000). While ethical codes of practice are critically important, the bigger challenge is to create an environment that facilitates the operationalisation of these principles. Our central concern, therefore, is the need to move beyond the emphasis on harm avoidance in the institutional and professional ethics of contemporary cities towards a new ethical paradigm shaped by deliberate measures addressing inequality, governance, democracy, social inclusion and environmental sustainability. This is of course ultimately a political project. It takes place within political environments with various stakeholders (elected representatives, business interests, community groups, etc.) applying pressure under the radar on individual planners and where the increasing privatisation of the public sphere creates new ethical dangers.

Contemporary neoliberal urbanism, as discussed in Chapter 1, has its own ethical frame that manifestly impacts on forms of political, institutional and personal power. From the perspective of normative ethics, neoliberalism proposes economic growth via market mechanisms as the priority for wealth creation with limited concern for the negative impacts on individual welfare, society and the environment (Astroulakis 2014). A core goal of neoliberalism is to increase freedoms and liberate individuals from collective social responsibility (Benatar et al. 2018). The burden of responsibility for social ills is thus shifted so that each person individually is responsible to somehow address the instances where capitalism morally and structurally fails (Bloom 2017, p. 17). The truth is that we are not becoming individually less ethical and less moral. Nor are we intimating that as individuals in a neoliberal system we are all committed to profit over people and the planet (Elkington 2018). Rather, two interconnected developments have occurred. First, the terrain of the public realm has changed; public concerns are increasingly marketised rather than regulated (think of Silicon Valley self-regulating on privacy rights and AI) and thus eliminating the opportunity (some would say need) for legal, political and ethical interference (Brown 2015, p. 140). Second, the burden of responsibility to make society, workplaces and everyday lives more ethical in the face of a permanent free market lies firmly on the shoulders of ethical individuals who are in turn operating in an increasingly constrained public space (Bloom 2017). Forces of capital overlay contemporary patterns of urban governance in a uniform, global and omnipresent manner, but this results in a very uneven landscape of change and inequality. Nevertheless, as we discuss in Chapter 8, there are numerous examples of urban communities pushing back against the neoliberal project, particularly as it enters an increasingly volatile phase. The opportunity exists to elaborate on a new era of urban ethics.

Rebuilding on ethical foundations

Rebuilding the city along ethical lines requires, as a first step, ethical foundations. Just as under neoliberalism we find the ethical burden is focussed upon the individual, so the challenge is to rediscover *collective* ways of moving along ethically guided pathways towards a sustainable future. There are diverse ways of describing, categorising, structuring and conducting ethical practice but, in this chapter, we follow convention by dividing them into normative and non-normative streams. Normative ethics focuses on human conduct and seeks to address the kinds of conduct that are considered good, bad, right and wrong; that is, *how conduct should be guided*. This implies consideration of what urbanites and their communities *ought to do* if they are to attain the ideal or ethical city. Non-normative ethics, which includes meta-ethics and descriptive ethics, involves a study of the nature and methodology of moral judgements such as: What do people and communities posit as ideal or good?; How are moral beliefs justified?; What proportion of the population agree with particular carbon abatement strategies, urban design propositions or ways of managing household waste? Table 3.1 presents the sub-categories of normative and non-normative ethics: consequentialism, deontology, virtue ethics, descriptive ethics and meta-ethics. In doing so, it provides details of the main principles, philosophical underpinnings, proponents and characteristics of each.

TABLE 3.1 Understanding normative and non-normative ethics

Main Categories	Normative Ethics			Non-Normative Ethics	
	Consequentialism (Teleological end, goal, purpose, outcome dependent)	*Deontology* (Duty, obligation, principle)	*Virtue Ethics*	*Descriptive Ethics*	*Meta-ethics*
Ethical Focus	Conduct (What should we accept and why?)		Character (What is the most appropriate behaviour?)	Propensity, Comparison	Understand nature of ethical properties, statements, attitudes and judgements
Proponents	Bentham, Mill: Utilitarianism. Singe: Preference Utilitarianism. Egoism (Personal ethics)	Kant: Categorical imperative – universal law, humane treatment, rationality. Moral absolutism	Socrates: Virtue ethics. Aristotle: Moral theory. Virtues of mind and character. Culturally relative	Study of actual conduct and notions of morality	Analysis of language, concepts and methods of reasoning. Study of ethical terms
Abstract Description	Action is right if it promotes the best consequences	Action is right if it is in accordance with a moral rule or principle	Action is right if it is what a virtuous agent would do in the circumstances	Examines what people believe to be right and wrong	Explores moral semantics, ontology and epistemology
Specification	The best consequences are those in which happiness is maximised (and misery minimised)	A moral rule is one that is required by rationality	A virtuous agent is one who acts through virtuous character traits that human beings need to flourish and live well	Provides the contextual understanding of how right and wrong are interpreted, without judgement	What is the meaning of moral terms? What is the nature of moral judgements? How are they supported?

(Continued)

(Continued)

Main Categories	Normative Ethics			Non-Normative Ethics	
	Consequentialism (Teleological end, goal, purpose, outcome dependent)	Deontology (Duty, obligation, principle)	Virtue Ethics	Descriptive Ethics	Meta-ethics
Characteristics	Critical reasoning and consideration of all possible ramifications of actions before making decision	Consistency, human dignity, universality, divine command, social conventions, social contracts	Aristotle's 18 virtues – including courage, truthfulness, modesty, friendliness, liberality, etc.	How many people think it is acceptable to give or receive a bribe?	What do we mean by the just city? What does good mean? What does right mean?
Real-world manifestation	Sustainable development goals, sustainability, triple bottom line, low carbon, Green New Deal, resilience, good governance, equitable society, social inclusion, democracy, poverty elimination,	Code of ethics (institutional) Civic duties and responsibilities Global Compact Principles The City we Need Principles	Characteristics of an ethical leader, ethical citizen, ethical consumer, whistle-blower, activist Professional codes of practice (individual)	Do people vote? Do people participate in community affairs? Forms of discrimination Are people corrupt?	What is the punishment for ethical breaches? What is a conflict of interest?

The central question addressed in this book is how can ethics act as a means to positively transform our urban way of life? We acknowledge that people have always understood the importance of ethics and civic duty. This is the base from which to seek ethical urbanism. Moreover, notions of ethics and civic duty give impetus to the sociocultural transformations required to address the range of contemporary challenges that would otherwise waste vital resources, exacerbate inequality, cripple good governance, stymie sustainability and slow our efforts to address climate change. The ethical city is a space that draws upon experience from existing urban practices (descriptive ethics) while at the same time reflecting principled and ethical approaches (normative ethics) to leadership, governance, planning, economic development, sustainability and public engagement. Ethical urbanism can illustrate potential pathways forward in the face of major disruptive forces by enhancing community resilience, enriching human well-being and ensuring sustainability. This is challenging since it can be argued that some examples of ethical urbanism merely demonstrate what the philosopher Peter Singer describes as preference utilitarianism – actions that fulfil the greatest amount of personal and communal interest, as opposed to actions that generate the greatest benefit for all society (Singer 2011). In each case, any analysis of existing urban projects reveals new perspectives as well as contested views on whether any proposed action is ethical or not. For instance, legislating for a universal basic income (UBI) in cities is viewed by many as a radical solution to growing inequality while others see it as potentially undermining society and the positive aspects of a strong work ethic. Similarly, while climate science points to the need to rapidly decarbonise our economies, the absence of universal agreement is cited as a reason for inaction by those segments of society who have a vested interest in the status quo.

Active dialogue on such ethical dilemmas helps make the moral beliefs underpinning contemporary practices explicit. Such discourse reminds us how cities operate and what and how this can change. The Big Leeds Climate Conversation is a good example of the in-depth negotiations needed. This four-month consultation (July to November 2019) was organised by the Leeds Climate Commission and set out to explore opportunities for change. The aim was to raise awareness of the need to tackle climate change, discover how residents perceived a number of bold ideas to cut emissions and examine the potential for transformative action. The ethical principles that arise from community dialogues such as this one in Leeds cannot be dictated by individual, partial or sectional group interests. These principles may reflect a universal viewpoint while also recognising that not all ethical judgements need be universally applicable (for instance, different cities and communities will adopt different approaches based on local circumstances). However, in making ethical judgements, we must collectively look towards decisions and actions that have the best consequences, on balance, for all affected. This also implies, metaphorically, that everyone must be willing to walk a mile in the shoes of those who would be worst affected. Deliberations around these questions will involve a degree of disharmony and discontent, since ethical questions are invariably shaped by conflicting rationalities. However, as we mentioned in Chapter 1, agonistic exchanges are welcomed since only by bringing into the open and challenging existing unspoken moral beliefs can we search for consensus. Embarking on such work is eased by a more overt consideration of normative and non-normative ethics and an understanding of how they relate to a meta-framework for an ethical city.

Normative ethics

Normative ethics concerns the principles or norms we ought to live by and addresses four main questions: What actions ought we do?; What ends are intrinsically good?; What character traits are virtuous? and What moral rights do we have? (Kagan 1997; Gensler 2018). We use normative ethical theories as a basis for arguing about what the right thing to do is in a given situation. For example, when facing an existential threat such as climate change or future pandemics, it can be argued that the right thing for people to do is to take action at a level that corresponds to that threat – there is a moral obligation to do so.

In many cities reducing carbon emissions and regaining local control of energy production has apparently been understood as the right thing to do (even though it is immensely problematic and progress often slow). For instance, as we will discuss further in Chapter 8, in 2011 the city of Berlin committed to reduce greenhouse gas emissions by 85 percent by the year 2050. When introducing the plan, the Berlin Senator for Urban Development and Environment, Michael Müller, also called for the re-municipalising of local energy production and energy supply networks as a key step towards carbon neutrality for the city. In other cities, there are ambitions borne of moral obligation that couple measures to reduce emissions with those designed to tackle urban inequality, to reduce the presence of cars and to increase the provision of more affordable housing. Cross-sectoral as well as multi-objective initiatives such as these offer increased potential to provide ethical city solutions. Oslo, for example, banned private cars from parts of the city centre in 2019 in a bid to heavily reduce greenhouse gas emissions even though this met with considerable push back (Williams 2019).

In the case of New York, also examined further in Chapter 8, the 2015 One New York City (One NYC) plan represents a form of social compact designed to deliver a Green New Deal that will create well-paid jobs, ensure equitable access to nature and guarantee the right to quality health care and education whilst promoting justice for ethnic minorities and other marginalised communities in the city. Applying a consequentialist ethics frame, the OneNYC 2050 plan postulates that inaction on climate change would have devastating consequences, with disproportionate harm on the most vulnerable groups. Moreover, the plan pursues a more inclusive economy because 'economic security and dignity are essential to overcome long-standing inequities' and for the creation of a city where everyone can succeed. The pursuit of healthy lives and the elimination of inequities in access to healthcare is a critically important goal recognising that 'healthcare is a human right' (New York City 2019, p. 14). Significantly, the advent of Covid-19 focused attention on the vulnerability of cities and their health systems to infectious disease outbreaks. As more research becomes available it is likely to expose how cities with high levels of inequality suffered severely with poor neighbourhoods, communities and individuals particularly affected (Bergamini 2020; Khan 2020). It would also highlight the central role cities play in being prepared for, mitigating and adapting to the consequences of pandemics (Muggah and Katz 2020). Consequentialism is one of three core theories in normative ethics, the other two being deontology and virtue ethics.

Consequentialism

Consequentialism (also referred to as *teleology*) implies that the consequences of the conduct of institutions, communities, groups and individuals should be the basis for any judgement about the rightness or wrongness. It proposes, rather simplistically, that an act is morally right only if it produces good outcomes or the best consequences (Kagan 1997, p. 61). While there are numerous variants, including two-level, motive and negative consequentialism, utilitarianism is currently the most influential consequentialist theory and will be our focus here. Utilitarianism holds that the most ethical choice is the one that will produce the greatest good for the greatest number of people – chiming with neo-classical economics in arguing for conduct to be assessed by its outcome or end result – its *telos*. The idea that decisions should pursue the greatest happiness for the greatest number of people is compelling, at least on the surface. It leads to discussions about principles, preferences, qualities and hierarchies of happiness and pleasure (Forcehimes and Semra 2019). According to John Stuart Mill: '[O]f two pleasures, if there be one to which all or almost all who have experience of both give a decided preference, irrespective of any feeling of moral obligation to prefer it, that is the more desirable pleasure' (Mill 2007, p. 286). Singer uses the term 'preference utilitarianism' to refer to maximising the satisfaction of preferences in this way (Singer 2011).

Urban participatory budgeting provides an excellent example of preference utilitarianism in action where people are directly involved in expressing their desires for how local funds should be invested. Introduced in Porto Alegre in Brazil in the late 1980s and since adopted in New York, Melbourne, Chicago and Madrid, this practice aligns municipal democracy with consequentialist theory. Municipal democracy in turn requires that civic leaders develop trust and respect from urban residents and prioritise inclusion as a prerequisite for budgeting. Further, the active pursuit of the 'greatest happiness' for the many through techniques such as participatory budgeting aims for an accountable, transparent and trusting public engagement process (World Bank 2008). This stands in sharp contrast to the reality of periodic elections and the tokenistic consultation found in many cities.

While utilitarian consequentialism provides useful tools for moral thinking and practice, it can leave large holes in ethical decision-making (Driver 2006; Gensler 2018). For example, its focus on individual preferences means that utilitarianism does little to tackle inequality and leaves little room for questions outside the pleasure-maximising self, such as universal human rights, inclusive ethics and intrinsic rights of non-human nature (Persson 2017, pp. 137–145). The complex questions of species rights, climate change, population limits, etc. point to broader ethical concerns around equity, human rights and ecological prudence. None of these has simple answers (O'Neill 2008). Refinements proposed to address such shortcomings include 'rule consequentialism/utilitarianism' (do what would be prescribed by rules with the best consequences) and 'act utilitarianism' (do the act with the best consequences) (Hooker 2005; Frey 2013). These essentially attempt to reconcile consequentialism and deontology, as discussed below.

Deontological ethics (non-consequentialism)

The non-prescriptive nature of consequentialism contrasts with deontological ethics, which is derived from the Greek word *deon*, meaning duty or obligation. This is sometimes referred to as rule-based ethics, where rules play a role in binding an individual or a group to a duty. Known also as non-consequentialism, this ethical theory holds that some actions, such as lying, murdering the innocent, torturing prisoners and keeping slaves are wrong in and of themselves for contemporary society irrespective of culture (McNaughton and Rawling 2006, p. 425). While the consequences of these actions are also considered, this does not extend to the primacy of maximising good for all as in consequentialism (Kamm 2013). There are four component parts of deontological ethics – duty, virtues, commandments and rights. All four are based on practical reasoning around our biological nature, pluralistic views on intrinsic values (as we discussed in Chapter 1) and the need for some kind of strict rules (Gensler 2018, pp. 193–210). In Austria, for example, there has long been sufficient political agreement that society should be responsible/obligated to ensure an adequate supply of social housing. Indeed, housing can be understood as a basic human right (Reinprecht 2007). Accordingly, the Vienna city government sees it as a duty to provide housing for every resident. This contrasts sharply with the situation facing cities in the UK where, since the 1980s, the emphasis has been on the 'right to buy' resulting in the sale of social housing and an endemic affordable housing crisis.

Deontological ethics is thus about rule-based rights and is usually represented in local government ethic codes and guidelines. It is the duty of local government officials to be a servant of the public, to act in the public interest and maintain public trust, to be accountable, to not seek personal benefit from their office and to be fair and impartial. This is essentially a Kantian framing around duty rather than ends, arguing that it is not the consequences of an action that make it right or wrong but the motivations of those who carry out that action. While this sounds straightforward in theory, reality is much more complex. Suppose Person A is running for mayor and has a plan to greatly benefit their urban community, but he will only be elected if they lie about their opponent. Should they lie? Consequentialists might say yes, but non-consequentialists (who believe it is wrong to lie regardless of the consequences) would say no. The other consequence we may wish to avoid, however, is the perpetuation of dirty politics where the best liars get elected (modified from Gensler 2018, p. 185).

Here is another example. Suppose a city has an acute shortage of affordable homes and unless this problem is remedied quickly the number of homeless will increase dramatically. In the city, however, there are thousands of unoccupied homes owned by banks that were repossessed from local people who defaulted on their mortgage payments following 2008 GFC. The banks do not want to rent out the homes, they would prefer to sell them. Should the city government appropriate these properties in order to ameliorate the shortage of social housing? Or do the banks have the right to leave the properties empty even though this course of action is worsening the situation in the local housing market? This was the dilemma facing the city of Barcelona, where the decision was made to expropriate a small number of houses to city control. This transfer could be avoided in cases where the bank found a tenant within three months paying a city-adjudicated social rent. (O'Sullivan 2018). It is a very controversial policy raising clear concerns about who is right and wrong, who sets the rules and for whom. A consequentialist might argue that the city is right to act because such measures are required to address the acute housing shortage and to provide access to affordable housing for people who might otherwise become homeless. A deontological view might also imply that the city has a duty to act even though such an intervention contravenes agreed rules and the rights of the banks. On the other hand, the ethical course of action for banks, from a deontological perspective, is compelled by their duty to answer to their shareholders and customers.

There may be occasions where we act with the right motivations (say to protect vulnerable members of a community) only to unintentionally worsen their plight. For example, decisions were taken in San Diego in the 1980s to concentrate service providers for the city's homeless in the East Village area. The downside of this decision, formulated with the best intentions, was the creation of a highly visible concentration of homeless people – equivalent to a ghetto – with some residents expressing concerns about local safety in the aftermath (Halverstadt 2016). A response to the unintended consequences can be found in Kant's concept of the categorical imperative. At the risk of oversimplifying what is a rather complex idea, this requires conformity with moral rules, universal laws, and act in a rational manner in the treatment of others. Specifically, the categorical imperative necessitates that an individual 'act only in accordance with the maxim through which you can at the same time will that it become a universal law' (Kant 2002, p. 39). This implies appreciation of how the desired end could be undermined or frustrated if the action were universalised. Following the categorical imperative would require avoidance of decisions that lead to the creation of homeless ghettos in cities since that may be universalising an insensitive attitude towards the plight of the homeless and their host communities. If all homeless are concentrated in single locations within the city, what impact would this have? A categorical imperative lacks contingent conditions: it has to be shaped by both universality and impartiality.

Decisions in the ethical city, which recognise Kant's categorical imperative, establish rules for all to follow in the same situation. Decision-makers should reflect upon the universality and impartiality of a course of action. For instance, a requirement that all new housing developments within a city should include car parking may not align with the categorical imperative. While this rule may make sense in a city like Melbourne where only 10 percent of dwellings do not own a car, it makes less sense in Osaka where only 40 percent of people own a car, the second lowest in Japan. On the other hand, a requirement for every household to reduce its carbon footprint may well do so. Interestingly, these two requirements played out in relation to the Commons and the Nightingale housing development projects in Melbourne (Feagins 2018). Designed to meet the triple-bottom line of affordability, sustainability (including carbon neutrality) and sociability, these housing projects kept hitting a snag when it came to obtaining planning permission due to the fact that no car parking spaces were provided. The developers had to appeal to the Victorian Civil and Administrative Tribunal (VCAT) explaining that the apartment residents did not aspire to own a car and could rely on excellent, already existing public transportation connections (Edmunds 2017). Both the Commons and Nightingale projects have become recognised as leading exemplars of sustainable design across Australia and beyond (Moore and Doyon 2018).

Virtue ethics

Virtue ethics derive from Ancient Greek philosophy and particularly the works of Socrates, Aristotle and Plato. In the *Republic*, Plato presented Four Cardinal Virtues – wisdom, justice, fortitude and temperance. These are traits so well-entrenched in a person they become second nature. As such, to Plato, morality is not just about making good choices and acting on them, it is also about becoming a good person (Gensler 2018, pp. 31–33). To exhibit a virtue is to be a certain type of person with a particular mindset (Hursthouse 2001). A virtue is a disposition to act in certain ways and not in others (Annas 2006, p. 515). In the contemporary context, the notion of virtue ethics influences what we consider to be the characteristics of ethical leadership (Price 2008) or the traits moulding ethical citizenship or ethical consumerism. Virtue ethics, however, also embodies cultural relativity since different people, cultures and societies often have distinct opinions on what constitutes a virtue. So, for example, a virtuous leader who extols financialisation would have very different traits to a communitarian.

Ethical urban behaviours are thus shaped through a variety of factors, including material settings, cultures and philosophical orientations among them. They have two essential facets according to Varda (1999). The first is grounded in judgements and justifications. The second is reflexive and provides thoughtful responses to an incident, situation or decision. Reflexivity points to the differences between 'know-how' and 'know-what', i.e. between spontaneous coping and rational judgement. Ethical know-how suggests that there is a form of active and engaged ethics based on a tradition that identifies the good and virtuous. We may act on it with little thinking in situations as diverse as giving up a seat on the bus for an elderly person to risking one's life to save a drowning person. The know-what approach to ethics focuses on what is the right thing to do while the know-how approach focuses on what it is good to be (for instance the kind of leader or person you should or would like to be). Both affect how we live in an ethical city.

Moral principles (about actions) and virtues (character traits) intersect. What we ought to do in a situation is what a virtuous person who understands the situation and acts in character would do (Gensler 2018, p. 198). Moreover, a virtuous person is someone who has internalised the correct principles about how we ought to live. With reference to collective responsibilities, rather than those of the individual, it is possible to imagine communities that exemplify certain intrinsic virtues and values that align with specific principles (e.g. broadmindedness, a world at peace, equality, protecting the environment, social justice, helpfulness, forgiveness, honesty, self-acceptance, responsibility) as discussed in Chapter 1 (Crompton and Weinstein 2015).

Non-normative ethics

Descriptive ethics (people's beliefs about morality) and meta-ethics (ethical terms and theories) are two forms of non-normative ethics. Together, they provide a window into how things are, as a prelude to working out how they ought to be. This is a precursor to considering how we can bridge the gap between what ought to be (normative ethics) and what is (non-normative ethics). For example, in our MOOC on ethical cities, we asked participants whether or not it is justifiable for someone to accept a bribe. Eighty percent responded that this would never be acceptable, but the remaining 20 percent disagreed and, in some instances, argued that it was simply business as usual. That is, in some parts of the world if you want to get things done, you just have to pay a backhander. There are multiple layers to be peeled back here on how people interpret acts of bribery, the varieties they may take, the motivations for different forms and their consequences. Non-normative ethics helps with the identification of pressures and influences that impact on our ability to make objective decisions. When examining the actions of stakeholders within the city, we can examine how influence is bought, the kinds of political pressures exerted on local officials and the probity checks in place to counter these (or the lack thereof) (Cook and Sarkissian 2000). By surfacing such issues, it may then be possible to develop measures to promote and refine ethical practices, including codes of conduct, institutional mission statements and clear public policy statements around transparency, accountability and good governance.

Descriptive ethics (or comparative ethics as it is sometimes called) is empirically oriented, directed at understanding the ethical ideals held by individuals and groups in society and how they perceive complex ethical concerns. In the MOOC, we asked participants about what they considered to be the right thing to do in terms of measures that might help promote an ethical city. Figure 3.1 shows that close to 68 percent supported the idea of a Universal Basic Income (UBI), 85 percent ethical urban development, with 71.6 percent and 85.2 percent endorsing the need for more ethical business practices and market interference, respectively. The sharing economy and retaining local economic autonomy (i.e. keeping money in the local economy) both garnered support although the idea of adopting a local currency was less popular (only 36.2 percent agreed). While this discontent with 'business as usual' is expected from a self-selected sample of individuals who chose to study a course on ethical cities, it does reveal boundaries and relationalities in our thinking of what is possible within the constraints of any given situation. It is intriguing to note how these boundaries may have shifted as a result of the pandemic with, for instance, Spain becoming the first nation to commit to a permanent UBI in order to counter the economic impacts of the pandemic (Slater 2020). Thus, descriptive ethics is not static, nor is it limited to representing a flat view of the current status quo. It embodies reasoning, context and relationality and reveals opinions tied into moral reasoning.

Ethics is bounded by normative practice, logics of moral claims and arguments and of course morality itself, as the body of standards or codes derived from particular philosophical, religious or cultural principles. Thus, meta-ethics is primarily concerned with the analysis of language, concepts and methods of reasoning. It is primarily about trying to take a bird's-eye view and to understand the practice of ethics (Fisher 2014). Importantly, this extends to questions of how we know what is good or bad, what is a good or a bad result, what is reality and who decides such questions. In a brilliant sketch called the New Sliding Scale of Ethics, by two Australian satirists, Bryan Dawe is interviewing John Clarke. At one point, Clarke says, 'If you did a bad thing in order to get a result, and you got a result, then you haven't done a bad thing.'

Dawe looks a bit perplexed and asks, 'Didn't you just say that you did a bad thing?'

'No,' continues Clarke. 'Not if you got the result you intended to get.'

Dawe responds, 'Well, who says you got a result?'

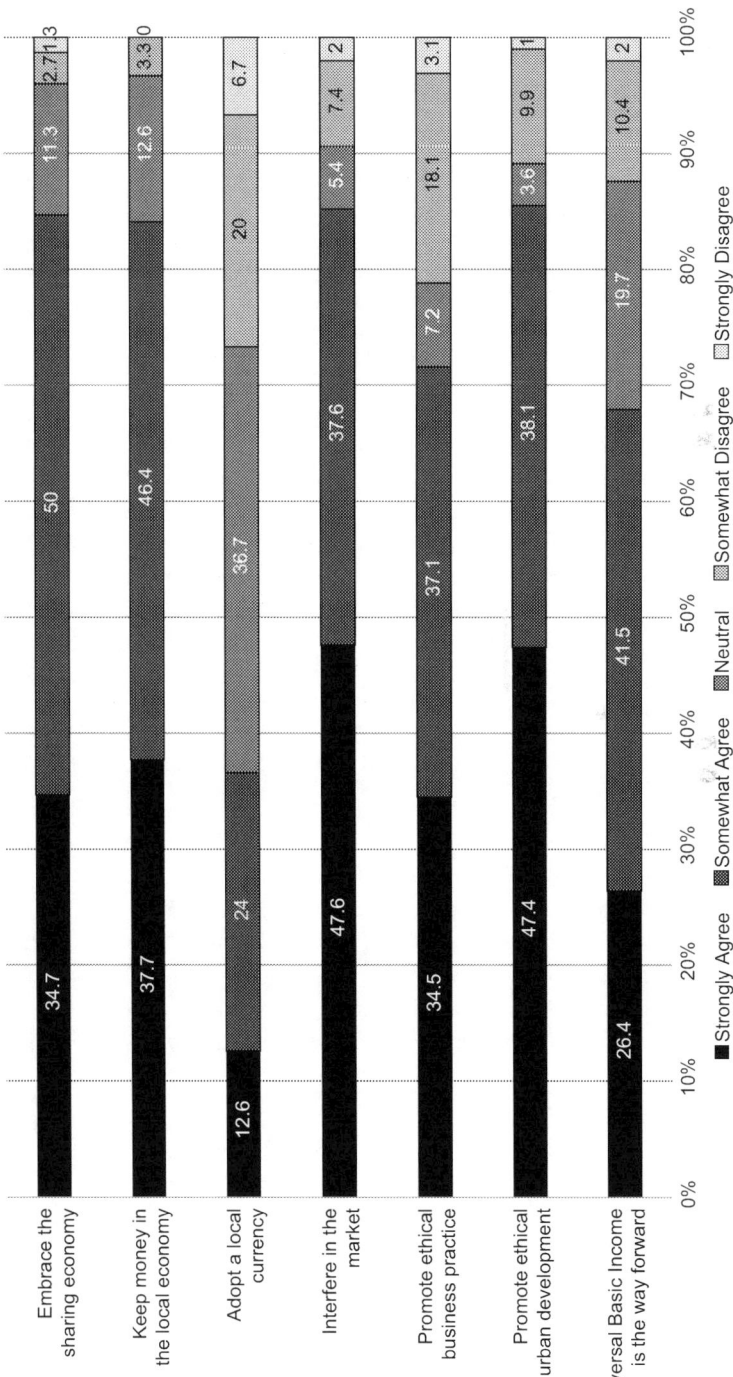

FIGURE 3.1 What is the right thing to do?

Source: Survey of participants (*n* = 150 to 190) in the Ethical Cities MOOC 2016–2017.

Clarke answers, 'You do. You would need to establish that.' Clarke then explains that what he is talking about is called 'ethical standards in the context of reality'.

'And who defines reality?' asks Dawe.

'You do!' insists Clarke (Clarke and Dawe 2015).

According to Peter Singer, many people mistakenly consider that what is good and bad is relative to the society or culture in which they live, or is merely a matter of subjective taste or opinion (Singer 2011). This is ethics as cultural relativism (right and wrong depends on the place and/or time), subjective relativism or situational ethics (i.e. there are no universal moral norms for right or wrong, only personal subjective decisions) or supernaturalism (judgements are based upon God's will). These so-called popular meta-ethics imply that what is good is socially approved somehow in a given culture (Gensler 2018). In philosophical meta-ethics, such social self-referencing meets intuitionism, emotivism and prescriptivism (van Roojen 2015). Intuitive ethics relies on ideas of truth and moral principles based on intuition, whereas emotivism or ethical nihilism speaks to life without objective meaning, purpose or intrinsic value and therefore without moral truths, facts and knowledge. The contemporary rise of nihilism, fatalism and resentment associated with a lack of moral compass can thus be traced to a late neoliberalism that espouses 'freedom unbridled and uncultured, freedom to put a stick in the eye of accepted norms, freedom from care of the morrow' (Brown 2019, p. 171).

Is it too late?

Although moral philosophy is largely esoteric, ethical frames are never lacking (Persson 2017, p. 22). What lacks is awareness and overt attention to the consequences and principles playing out. In the absence of attention to enactment of teleology and deontology, ethical notions around cultural relativism, subjectivism and emotivism insert themselves in the way of progress, causing dramatic regress. Over the past four decades we have witnessed growing inequality, the privatisation of public infrastructure and services, the crushing of labour unions, deindustrialisation of Western societies, rapid financialisation and the rise of a super-rich '1 percent' class, alongside the rise of private economic power. For the other 99 percent, the reality is longer working hours for lower relative wages and a preoccupation with economic survival. Under these circumstances, today's needs understandably take precedence over other considerations about the longer-term future. If you are living from pay cheque to pay cheque, just getting by leaves little time to determine what is right and wrong according to social norms and codes. This may especially be the case when questions of ethics are apparently resolved by neoliberal economic realities and rules, and there is no recourse to reopen them. Worsening inequality fosters greater dissonance and lower engagement, contributing to the weakening of shared ethics. We suspect, however, that without a continuously attended to and maintained ethical practice in cities, social and ecological misery is an imminent prospect for many.

An ethical pathway out of this situation implies a means can be found through which a push for more empathetic, caring, just and engaged urban ways of living becomes realistic. Surely such efforts will face intense opposition from those who like things just the way they are, who view these ethical ideas as irrelevant and disruptive to well-established behavioural habits. It may also be unrealistic to think of the ethical city as some sort of universal remedy to the numerous ills that face society today. The ethical city is a noble objective, but what else? The ethical frame provides a way to provoke new exchanges and imaginaries. It emphasises transparency and provides an opportunity to overcome conflicts by acknowledging co-dependency. In the ethical city social relations, dignity, accountability and consciousness are pushed centre-stage. By such means, ethics can be understood as 'a powerful lever for transformation, one that has as its goal learning how to be and how to live together' (Langlois 2011, p. 97). It is never too late and seismic shifts in ethical practice can be profound and rapid.

In this era of constant crises, it is not only judicious to expedite ethical city practice, it is also necessary, as other nihilistic ethics are fighting for supremacy in the future of cities. What should be done as a matter of principle and what should be allowed to all urban residents as a matter of right? While answering this might reveal a wide range of possibilities, the critical point is that the questions are asked. Moreover, it is hard to imagine how our current predicament of inequality and ecological crisis coupled with social and political dissonance would have become so pressing if empathy, care, justice and critical engagement were prevalent as values in urban society. Hence, while these are not mutually exclusive nor all-encompassing, they serve to illustrate how intrinsic values are necessary in underpinning ethical city practice.

References

Annas, J. (2006) Virtue ethics, in Copp, D. (ed.) *The Oxford Handbook of Ethical Theory*. Oxford: Oxford University Press.

Astroulakis, N. (2014) An ethical analysis of neoliberal capitalism, *Alternative Perspectives from Development Ethics, Ethics and Economics 11* (2). [Online]. Available: https://core.ac.uk/download/pdf/55653674.pdf [Last accessed: 10 May 2020].

Benatar, S., Upshur, R. and Gill, S. (2018) Understanding the relationship between ethics, neoliberalism and power as a step towards improving the health of people and the planet, *The Anthropocene Review 5* (2): 155–176. https://doi.org/10.1177/2053019618760934.

Bergamini, E. (2020) How COVID-19 is laying bare inequality, *Bruegel*. [Online]. Available: www.bruegel.org/2020/03/how-covid-19-is-laying-bare-inequality/ [Last accessed: 13 April 2020].

Bloom, P. (2017) *The Ethics of Neoliberalism: The Business of Making Capitalism Moral*. Routledge Studies in Business Ethics. London and New York: Routledge.

Brown, W. (2015) *Undoing the Demos: Neoliberalism's Stealth Revolution*. New York: Zone Books.

Brown, W. (2019) *In the Ruins of Neoliberalism: The Rise of Antidemocratic Politics in the West.* New York: Columbia University Press.

Calvino, I. (1972) *Invisible Cities.* London: Vintage Books, published in 1997.

Clarke, J. and Dawe, B. (2015) *The New Sliding Scale of Ethics: Officially Launched this Week.* [Online]. Available: www.youtube.com/watch?v=O0l_UDUqpDE&feature=youtu.be [Last accessed: 2 April 2020].

Cook, A. and Sarkissian, W. (2000) Who cares? Australian planners and ethics, in Bishop, P. and Preston, N. (eds.) *Local Government, Public Enterprise and Ethics.* Sydney: The Federation Press.

Crompton, T. and Weinstein, N. (2015) Common cause communication: A toolkit for charities, *Common Cause Foundation.* [Online]. Available: Common cause communication: A toolkit for charities [Last accessed: 8 May 2020].

Das, V. and Randeria, S. (2015) Politics of the urban poor: Aesthetics, ethics, volatility, precarity, *Current Anthropology 56* (S11): S3–S14. https://doi.org/10.1086/682353.

Driver, J. (2006) *Ethics: The Fundamentals.* Malden, Oxford and Victoria: Wiley-Blackwell.

Edmunds, S. (2017) Nightingale 2: No parking OK with VCAT, *The Fifth Estate.* [Online]. Available: www.thefifthestate.com.au/innovation/residential-2/nightingale-2-no-parking-ok-with-vcat/ [Last accessed: 21 April 2020].

Elkington, J. (2018) 25 years ago I coined the phrase 'triple bottom line': Here's why it's time for a rethink, *Harvard Business Review.* [Online]. Available: https://hbr.org/2018/06/25-years-ago-i-coined-the-phrase-triple-bottom-line-heres-why-im-giving-up-on-it [Last accessed: 10 May 2020].

Feagins, L. (2018) Nightingale housing wants you to own a great apartment, *The Design Files.* [Online]. Available: https://thedesignfiles.net/2018/02/nightingale-housing-wants-you-to-have-a-nice-apartment [Last accessed: 3 July 2018].

Fisher, A. (2014) *Metaethics: An Introduction.* Abingdon and New York: Routledge.

Forcehimes, A.T. and Semrau, L. (2019) *Thinking Through Utilitarianism: A Guide to Contemporary Arguments.* Indiana: Hackett Publishing Company.

Frey, R.G. (2013) Act-Utilitarianism, in LaFollette, H. and Persson, I. (eds.) *The Blackwell Guide to Ethical Theory.* Oxford: Wiley-Blackwell. https://doi.org/10.1111/b.9780631201199.1999.00012.x.

Gensler, H.J. (2018) *Ethics: A Contemporary Introduction*, 3rd Edition. New York and London: Routledge.

Halverstadt, L. (2016) How Downtown San Diego's East village became the homeless ghetto, *KPBS.* [Online]. Available: www.kpbs.org/news/2016/aug/17/how-east-village-became-homeless-ghetto/ [Last accessed: 2 April 2020].

Hooker, B. (2005) Right, wrong, and rule-consequentialism, in West, H. (ed.) *Blackwell Guide to Mill's Utilitarianism.* Boston: Wiley-Blackwell.

Hursthouse, R. (2001) *On Virtue Ethics.* Oxford: Oxford University Press.

Jovanovic, S. and Wood, R.V. (2006) Communication ethics and ethical culture: A study of the ethics initiative in Denver city government, *Journal of Applied Communication Research 34*: 386–405. https://doi.org/10.1080/00909880600908633.

Kagan, S. (1997) *Normative Ethics.* New York and London: Routledge.

Kamm, F.M. (2013) Nonconsequentialism, in LaFollette, H. and Persson, I. (eds.) *The Blackwell Guide to Ethical Theory.* Oxford: Wiley-Blackwell. https://doi.org/10.1111/b.9780631201199.1999.00014.x.

Kant, I. (2002) *Groundwork for the Metaphysics of Morals.* Edited and Translated by A.W. Wood. New Haven and London: Yale University Press.

Khan, S. (2020) The BAME people are dying from coronavirus: We have to know why, *The Guardian.* [Online]. Available: www.theguardian.com/commentisfree/2020/apr/19/bame-dying-coronavirus-sadiq-khan [Last accessed: 21 April 2020].

Langlois, L. (2011) *The Anatomy of Ethical Leadership: To Lead Our Organizations in a Conscientious and Authentic Manner*. Edmonton: Athabasca University Press.

Los Angeles City (2019) *Ethics Handbook for City Officials*, Los Angeles City Ethics Commission. [Online]. Available: https://ethics.lacity.org/wp-content/uploads/2019/01/City-Officials-Handbook-2019-with-Cover-1.pdf [Last accessed: 10 May 2020].

McNaughton, D. and Rawling, P. (2006) Deontology, in Copp, D. (ed.) *The Oxford Handbook of Ethical Theory*. Oxford: Oxford University Press.

Mill, J.S. (2007) Hedonism, in Shafer-Landau, R. (ed.) *Ethical Theory: An Anthology*. Oxford: Blackwell Publishing.

Moore, T. and Doyon, A. (2018) The uncommon nightingale: Sustainable housing innovation in Australia, *Sustainability 10* (10): 3469. https://doi.org/10.3390/su10103469.

Moroni, S. (2013) Afterword: Ethical problems of contemporary cities, in Basta, C. and Moroni, S. (eds.) *Ethics, Design and Planning of the Built Environment, Urban and Landscape Perspectives*. Heidelberg, New York and London: Springer.

Mostafavi, M. (ed.) (2017) *Ethics of the Urban: The City and the Spaces of the Political*. Zurich: Lars Muller Publishers.

Muggah, R. (2014a) How to protect fast growing cities from failing. [Online]. Available: www.ted.com/talks/robert_muggah_how_to_protect_fast_growing_cities_from_failing [Last accessed: 10 May 2020].

Muggah, R. (2014b) Deconstructing the fragile city: Exploring insecurity, violence and resilience, *Environment & Urbanization 26* (2): 345–358. https://doi.org/10.1177/0956247814533627.

Muggah, R. (2016) How fragile are our cities? *World Economic Forum*. [Online]. Available: www.weforum.org/agenda/2016/02/how-fragile-are-our-cities [Last accessed: 27 April 2016].

Muggah, R. and Katz, R. (2020) How cities around the world are handling COVID-19: And why we need to measure their preparedness, *World Economic Forum*. [Online]. Available: www.weforum.org/agenda/2020/03/how-should-cities-prepare-for-coronavirus-pandemics/ [Last accessed: 13 April 2020].

New York City (2019) One NYC 2050: Building a strong and fair city, volume 1 of 9. [Online]. Available: https://onenyc.cityofnewyork.us/wp-content/uploads/2019/05/OneNYC-2050-Full-Report.pdf [Last accessed: 10 May 2020].

O'Neill, B.C. (2008) *Population and Climate Change*. Cambridge: Cambridge University Press.

O'Sullivan, F. (2018) In search of affordable housing, Barcelona turns to repossessed homes, *Citylab*. [Online]. Available: www.citylab.com/equity/2018/04/barcelona-is-taking-over-repossessed-homes/558239/ [Last accessed 13 April 2020].

Persson, I. (2017) *Inclusive Ethics: Extending Beneficence and Egalitarian Justice*. Oxford: Oxford University Press.

Price, T.L. (2008) *Leadership Ethics: Introduction*. Cambridge: Cambridge University Press.

Reinprecht, C. (2007) Social housing in Austria, in Whitehead, C. and Scanlon, K. (eds.) *Social Housing in Europe*. London: London School of Economics.

Singer, P. (2011) *Practical Ethics*, 3rd Edition. Cambridge: Cambridge University Press.

Slater, N. (2020) Spain's UBI is a wake-up call for Americans, *Current Affairs*. [Online]. Available: https://www.currentaffairs.org/2020/04/spains-ubi-is-a-wake-up-call-for-americans [Last accessed: 13 April 2020].

Standing, G. (2011) *The Precariat – The New Dangerous Class*. London, New York, Oxford, New Delhi, Sydney: Bloomsbury Academic.

Thompson, W.N. and Leidlein, J.E. (2008) *Ethics in City Hall: Discussion and Analysis for Public Administration*. Sudbury, Ontario and London: Jones & Bartlett Publishers.

van Roojen, M. (2015) *Metaethics – A Contemporary Introduction.* New York and Abingdon: Routledge.

Varda, F.J. (1999) *Ethical Know-How: Action, Wisdom, and Cognition.* Stanford: Stanford University Press.

Welch, D.D. (2014) *A Guide to Ethics and Public Policy: Finding Our Way.* New York and Abingdon: Routledge.

Williams, L. (2019) What happens when a city bans cars from its streets? *BBC.* [Online]. Available: www.bbc.com/future/article/20191011-what-happens-when-a-city-bans-car-from-its-streets, 16 October 2019 [Last accessed: 2 April 2020].

World Bank (2008) Brazil: Toward a more inclusive and effective participatory budget in Porto Alegre, volume 1. Main report. [Online]. Available: https://openknowledge.worldbank.org/handle/10986/8042 [Last accessed: 30 August 2016].

4

WHO SHAPES THE ETHICAL CITY?

If each city is like a game of chess, the day when I have learned the rules, I shall finally possess my empire.

Italo Calvino, *Invisible Cities*, 1972, p. 109

Re-configuring cities as ethical

Cities are shaped by people and their spatially defined communities. They can be understood materially as physical urban landscapes and sites of capital accumulation. In a normative sense, city shaping involves complex and continuous competition between diverse interest groups, ideas and ideologies. This uneven contest engages politicians, urban planners, architectural designers, engineers, investors, developers, civil society and communities in institutional and policy processes seeking to direct and mediate capital flows with divergent implications for urban futures. City shaping is a matter of coalescing conflicting priorities around economic development, sustainability, mobility, liveability, community needs, cultural development, housing, health and security (Rapoport et al. 2019). The robustness of negotiated outcomes is influenced by a range of factors including vibrancy of local democratic practice, levels of political accountability, degrees of process transparency and respect for human rights. Ethical principles, rights and the search for shared values around common good, as we discussed in the preceding chapters, can also be influential in shaping these outcomes.

As cities are manipulated by people and interests in particular ways, so neoliberal governance has heralded new public management. Law is replaced with self-governing benchmarking, shifting the balance of conflicting interests with stakeholders and perpetuating political and normative challenges as having inevitable and unquestioned technical, optimising and best practice solutions (Brown 2015, p. 71). Terms such as 'stakeholder' indeed conform with neoliberal language conveying an objectivity and flat arena of power and accountability that of course does not exist. Laudable aims to community cohesion, social inclusion, respect, care and justice for people and planet are inevitably conflated within often subtle as well as not-so-subtle fault lines of (dis)advantage. We now face a situation where 'citizen virtue is reworked as responsibilized entrepreneurialism and self-investment, it is also reworked in the austerity era as shared sacrifice' (Brown 2015, p. 210). Recognition of who is actually shaping events and in whose interests is a precursor to contemplating the ethical city reshaped so that direct democratic accountability and questions of sustainability are at the heart of urban decision-making, design and functioning.

In an early version of our Ethical Cities MOOC we divided conflicting interests for reasons of simplicity into ethical leaders, ethical planners, entrepreneurs and ethical urbanites (see Table 4.1). In the ethical city, questions of accountability and representation are taken as given, and so shaping the city becomes centred upon the leadership of ideas rather than the pursuit of divisive interests or the notion of the ethical leader.

Ethical leadership of ideas

Leadership is not simply a matter of style or popularity; it is about ideas and values and involves accountable, authentic, democratic and just action to bring about positive change in society (Fien and Wilson 2015). This encompasses understanding of diverse and sometimes conflicting societal needs and energising people to pursue a better state of affairs than they had ever thought possible. Leaderships is about creating a values-based umbrella large enough to accommodate diverse interests in society but focused enough to direct our energies to the pursuit of the greater good (O'Toole 1996). Leadership ethics itself is a complex topic with moral theories and empirical evidence assisting the evaluation of why leaders justify sometimes breaking the rules (Price 2008).

For Plato, the most appropriate people to lead in our cities are uninterested in money or prestige but inspired by higher motivations to only do things that are advantageous for the city and its citizens. However, the sad reality is that today 'many people are attracted to politics by a gnawing need to be loved and admired. Public service is not only an important role for such people, but it also fills a hole in their emotional needs' (Wechsler 2013, p. 963). This shortcoming is something that city administrations have sought to address from time to time. For example, the Barcelona en Comú (Barcelona in Common) platform promotes a policy agenda focused on social justice, community rights and participatory democracy and, in 2014, developed a code of political ethics that members keen to pursue political office or an official position must adopt. Based on the principle of governing by obeying, the code requires elected politicians be responsive to the people they represent, as illustrated in Table 4.2.

A basic requirement is that political representatives commit to the Universal Declaration of Human Rights. They must respect the right to freedom of opinion, of assembly and association, and the right to take part in government, to social

TABLE 4.1 Conflicting interests shaping the ethical city

Ethical Urban Leadership	Ethical Urbanists
Ethical urban leadership of ideas plays a critical role in reforming political processes, setting out ambitious visions (rather than defending the status quo) and tackling corrupt practices. While placing emphasis on the role of accountable and transparent leaders, the core concern is ideas and how they develop through constant dialogue with communities while leaders seek to maintain public trust and fairness. *Characterised by:* A clear code of ethics that ensures leadership is accessible to a widest spectrum of often conflicting ideas.	The agenda of the urbanists (includes planners, developers, architectural designers, etc.) is crowded and complex. It requires the ability to deliver on competing goals – sustainability, liveability, resilience, economic vibrancy and social inclusivity. Ultimately accountable to their community, ethical urbanists approach their work with humility and objectivity. *Characterised by:* Personal engagement with community, making explicit the values underpinning their planning, design and policy decisions. This requires a holistic vision, integrated approach and awareness of global implications.

Ethical Practitioners in Democratic Communities	Ethical Community Entrepreneurs
Starts with recognition of individual rights (International Declaration of Human Rights) and extends to civic duties and responsibilities. Residents are empowered, connected and engaged with the community around them. *Characterised by:* Residents committed to and valuing their role in shaping a city based on equity, accountability and sustainability.	Ethical community entrepreneurship focuses on local ownership of prosperity generating activities and assets. It promotes an ethical stance by local businesses influencing how they operate and interact with each other. Public anchor institutions, including city governments, support this by using their financial spending power to generate community wealth and fostering an economy that functions for all residents. *Characterised by the following:* Community wealth building, respect for labour rights (living wage), decent jobs and fair work, ethical procurement and supply chains, local enterprise and worker-owned cooperatives.

Source: Barrett et al. (2016).

security and to work. The same political representatives are required to resign when charged with crimes of racism, xenophobia, violence, homophobia or other crimes against human or labour rights. The code combines both rules-based deontological approaches (e.g. limits on campaign financing and donations, making a calendar of public meetings, etc.) with ends-based, teleological approaches (e.g. uphold the election manifesto, goal-based crowd funding, etc.). This contrasts with

TABLE 4.2 Governing by obeying, Barcelona en Comú Code of Political Ethics

---

Democratisation of Political Representation

- Uphold the election manifesto.
- Make public the representatives diary of appointments.
- Make public all income sources, wealth and capital gains.
- Make public the criteria used to select people appointed for public office.
- Be accountable to citizens in both assemblies and online.
- Accept the censure or dismissal of councillors for poor management or failure to implement the manifesto.
- Not to accept positions of responsibility within five years of leaving office in companies created or contracts implemented while in office.
- Keep regular contact with vulnerable groups.
- Guarantee citizen participation in relevant decision-making.

Financing and Transparency

- Be transparent in management of income and spending.
- Set maximum limits on private donations.
- Refuse to accept bank loans and donations that compromise political independence.
- Drastically limit election campaign spending.
- Promote goal-based crowd funding.
- Avoid use of foundations or entities that have different goals from the candidature.
- Make use of financing that is consistent with this code of ethics.

Professionalisation of Politics, Reduction of Privileges and Tackling Corruption

- Renounce gifts and privileges.
- Do not take on multiple public posts.
- Do not earn multiple salaries.
- Set maximum monthly salary to 2,200 Euros, including expenses.
- Limit term of office to two consecutive legislatures, extended to three in exceptional circumstances.
- Resign immediately if charged with crimes related to corruption, influence peddling, unjust enrichment, etc.

(Source: Barcelona en Comú 2014).

normative teleological approaches where, for local politicians, doing a good job means getting the best result for the greatest number of citizens. In some instances, how they get there matters far less than the result. In these cases,

> the process and the rules are things to be taken advantage of, to use, even abuse, in order to get the most for [their] constituents or to get the result [they] want. The ends justify the means and often determine what the means are in the first place.
>
> *(Wechsler 2013, p. 965)*

Something like the Barcelona en Comú code of political ethics cannot be expected to overcome normative leadership ethics and practices overnight but

TABLE 4.3 Limits and opportunities for ethical leadership in Barcelona

Limits	Opportunities
• Lack of a solid majority • Bureaucratic resistance • Financial limits and austerity imposed from upper tiers of government • Lack of powers in key policy areas • Global nature of economic and financial flows • Fierce opposition from pro-establishment coalition (economic elites, mass media, big national parties) • Difficulty in building a new political and cultural hegemony	• Maximising key institutional resources at the disposal of local governments • Confronting/putting pressure on upper tiers of government and international actors • Building political alliances with urban social movements and co-producing public policies • A multilevel struggle. Building a supralocal internationalist municipalism

Source: Blanco et al. (2019).

effective strategies for change are still possible. For example, elected representatives from Barcelona en Comú occupied only 11 of the 21 council seats in city hall between 2015 and 2019 (falling to 10 seats in the May 2019 elections), and so Barcelona en Comú chose to work as part of a coalition and was able to successfully nominate their candidate, Ada Colau, as mayor of the city. Barcelona en Comú had a number of advantages in effecting these strategies, including being rooted in local social movements and activism. It also benefits from Barcelona's history of critical engagement with innovative forms of representation and leadership. However, these advantages are, in turn, set against a broad array of significant obstacles and opportunities to change as illustrated in Table 4.3.

Ethical leadership exploits opportunities while circumnavigating limits. For example, upon taking office, Mayor Stephanie 'Steve' Chadwick of Rotorua, New Zealand (population 68,500) recognised the need to engage with the Indigenous Māori people of the Te Arawa in her community. Census data from 2013 shows that 34.3 percent of the local people identify as Māori, compared with 14 percent for New Zealand as a whole. Embracing the multicultural nature of the region, combined with acknowledgement of Māori culture and language, put Rotorua on a new path through the establishment of the Te Arawa Partnership Model. This resulted in a Sustainable Living Strategy, issued in July 2016, that is heavily influenced by and shows deep respect for Māori traditions. Remarking on this significant change in direction at the February 2016 Urban Thinkers Campus on Ethical Cities (in Melbourne), Mayor Chadwick explained that:

> the land our town is built on was gifted by the tribe, and so we've now developed a formal partnership model with them. That was hugely

> controversial, but we've got there and the sky hasn't fallen in. In fact, we'll
> be sitting around the council table on a much more inclusive platform now.
>
> *(RMIT University 2016)*

Re-elected in October 2019, Mayor Chadwick's platform involves driving forward a Vision 2030 to support community leadership, enhance the partnership with the Te Arawa and ensure local resilience, environmental sustainability, affordable housing and a vibrant city economy.

Both Barcelona and Rotorua are engaging with forms of ethical leadership that promote change through supralocal and international municipalism. In the case of Rotorua, the municipality signed up to the UN Global Compact Cities Programme in May 2015. Meanwhile, Barcelona hosted the International Municipalist Summit in June 2017 and participated in various international initiatives including UN-Habitat III and the Global Network of Cities, Local and Regional Governments. These horizontal networks of solidarity across cities play a key role in addressing complex issues that are too big for one city to handle. Moreover, in his manifesto on city leadership, the late Benjamin Barber argued that mayors and local executives often exhibit a non-partisan, pragmatic style of governance that is lacking in national and international halls of power (Barber 2013). Contrary to popular models of cities as competitive entities, he argued that cities are increasingly collaborating with each other, giving credence to ideas of democratic glocalism, of horizontalism rather than hierarchy and of pragmatic interdependence rather than outworn ideologies of national independence. In this optimistic scenario, cities and their leaders are projected as alternatives to nation states in tackling global problems, although many questions remain about the ability or likelihood of mayors to deliver ultimately on progressive agendas or on the extent to which the fate of cities remains tied to power structures in nation states. City autonomy is relative in a globalised market economy with the continuing ascendancy of nation states, and such initiatives depend upon cooperative actions, not only with other cities but also within multi-level systems of governance and across geopolitical regions.

Three ethical urban leadership characteristics have emerged through a review of the literature (Rapoport et al. 2019; Sennett 2018; Singer 2011; Langlois 2011). First, ethical urban leadership is an interactive and self-critical process requiring we live and act as one among many and be engaged by a world which does not mirror ourselves. Empathetic ethical urban leadership requires we put ourselves in the position of those affected by our decisions. Second, ethical urban leadership is not an individual activity but is distributed among many competing interest groups and institutions. Reaching consensus is no simple task since those affected by each and every decision can be extensive and diverse. The challenge, therefore, is to be able to identify group preferences and strive to take them into account. This and the frequency of engagement together indicate the quality of democratic representation processes. Third, ethical urban leadership recognises complexity, non-linearity and the need to reflect on outcomes beyond short political terms of office and political boundaries, to consider future generations and the future of the planet.

Ethical urbanists

In an address to the Congress of the International Society of City and Regional Planners, Wright (2013) asked whether we are all neoliberals now. The answer is yes, much like every citizen in the former Soviet Union was a communist. Successive waves of neoliberal thinking have pounded urban planning practice and practitioners into submission: resistance is futile in the face of the one omnipotent ideology. In a survey of the impact on neoliberalism on urban planning policies in the period from 1990 to 2010, Sager (2011: p. 179) identified four key impacts: (1) one-dimensional concentration on efficiency and economy in policy recommendations; (2) predilection for private, market-oriented solutions to urban problems, which favour property investors over community interests but which, incongruously, result in political compromises rather than economically sound decisions; (3) lack of a democratic agenda beyond consultation that elicits information from clients and consumers (which is how people are defined) and (4) indifference to concerns about unequal outcomes, exclusion, segregation and distributional questions. This is a world where private investors hold profitability as the highest priority (Sager (2011: p. 180)) and where urban planning systems have been transformed into 'quasi-market regulatory mechanisms' for mediating conflict (Healey 2000, p. 518). It has left planners, architects, mobility experts, developers and investors ever embroiled in a game of ethical trade-offs around urban development projects (Thompson and Leidlein 2008) and has resulted in land use zoning, urban regeneration and infrastructure provision often exacerbating social exclusion through gentrification, the loss of social housing and the increased privatisation of public space (Minton 2009; Nemeth and Hollander 2010; Sager 2011).

The broader impacts of neoliberalism on urban planning traditions are presented in Table 4.4, which also provides pointers to what we may expect next, especially around urban-led mobilisation in response to the climate emergency and other disruptive forces (Wright 2013; Rode 2019). Specifically, ethical leadership is posited as a counter to far-right nationalism and a platform for centre-left reformism and the progressive-left agenda (Stiglitz 2019). At one level, this recalls the rise of new urbanism and Jane Jacobs' battles with New York city planner, Robert Moses, invoking applied ethics in response to issues uniquely connected with urban place-making (Jacobs 1961; Kidder 2008). Jacob's four key contributions remain relevant today and include the importance of: (1) place, neighbourhoods and residents' experiences; (2) maintaining local sub-economies for people and neighbourhoods; (3) making visible that which is typically rendered invisible – people, places, neighbourhoods, heritage buildings, legacy and street scenes and (4) recognising the multiplicity of cultures and economies interwoven in our urban fabric (Sassen 2016).

TABLE 4.4 Ideological approaches and urban planning

Premodern	Modern	Postmodern	Neoliberal	Ethical Democracy
Planning Era				
Before World War I	1930s – avant garde movement 1940s to 1980s –planned economy	1960s to 1990s – counter culture 1980 – endorsed by the establishment	1970s to 2020s – neoliberal roll out globally	2010s onwards – scientific enlightenment and revolutionary reformism
Planning Theories				
Physical planning	Rational planning Systems planning Procedural planning	Advocacy planning Incremental planning Radical (action) planning Participatory planning Communicative planning	Strategic spatial planning	Localisation, Resilience planning Mitigation and adaptation
Humanistic Premise of Planning (Consequentialism)				
Utopia – An end state in which individuals are emancipated towards an ideal society	*Collective public interest* – An end state in which society en masse is emancipated towards a common good for the society	*Group interest* – An end state in which groups within society are emancipated towards a good defined by those groups	*Individual interest* – There is no end state for society or societal groups; but rather the right of each individual to pursue a good life that does not harm others	*Community as the highest standard* – Transition beyond the climate crisis as communities collaborate towards this end state

Epistemological Premise Underpinning Planning

Artistic design method – Physical and aesthetic design principles can be objectively defined by human reason	*Rational scientific method* – Planning principles can be defined through value-free scientific reason	*Participatory method* – Subjective value laden principles of individuals can be defined through participatory processes	*Managerialist method* – Individual good can be pursued through managerial process of defining and implementing goals, objectives and strategies	*Coping method* – Individual good is embedded in community good

Concept of the City

City beautiful – Cities are a symptom of social order and disorder	*City functional/mechanistic city* – Cities are an object that can be rational, ordered and mass produced	*Just city* – Cities are an expression of social diversity of its citizens and the ecological diversity of the environment	*Entrepreneurial competitive/ productive cities* – Cities are economic objects that are competing with each other for economic growth	*Ethical city* – cities are resilient, collaborative and mutually supportive in order to overcome disruptive technological, economic and sustainability challenges

Planning Governance

Limited uncoordinated community and government initiatives	Government-led with limited community involvement	Government-led with significant community involvement	Private sector-led through then market	Community-led with government and private sector engagement

Planning Approach

Government top-down with no bottom-up community involvement	Predominantly government top-down with some bottom-up community involvement	Predominantly bottom-up community involvement with top-down government involvement	Bottom-up through the market with limited top-down government engagement	Horizontal cross city community bottom-up involvement tied into international cooperation

(Continued)

TABLE 4.4 (Continue)

Planning Focus and Time Horizon

Premodern	Modern	Postmodern	Neoliberal	Ethical Democracy
Long-term planning at the city level (city visions) shaping physical urban form and aesthetic design	Medium-term planning at the city level (master blueprint and layout plans) focusing on detailed spatial urban form and infrastructure. Development control based on land use planning at the district level (zoning plans)	Medium-term infrastructure planning at the national and state levels. Detailed spatial urban form and infrastructure-based planning at the regional level (regional plans) or city level (master plans). Development control based on land use planning at the district level (zoning plans). Urban design-based planning at the local and site levels	Short-term spatial urban planning and infrastructure development at the city and district level in place of master plans (strategic spatial plans). Development-based planning at the local and site levels in place of development control-based zoning plans	Long-term plans for low-carbon infrastructure, retrofitting existing urban form. Resilience planning and risk management plans identifying vulnerabilities and hazard zoning. Sustainable design of new urban communities

Source: Wright (2013); Goodchild (1990).

Ethical urbanism is a strong thread throughout the work of David Harvey and others in the Right to the City movement (as elaborated in Chapter 2) who call for new and just ways of approaching together urban social processes and spatial form. Susan Fainstein, for example, argued that deliberative urban development strategies are critical in ensuring social inclusion, justice and care in cities (Fainstein 2010). Such concerns about how new forms of urbanism might emerge from balancing private profit-making and urban quality of life (liveability) provide insights into opportunities and limits as cities face a range of disruptive forces (Chapter 7). An agonistic approach to planning utilises climate crises, sustainability and liveability principles as a basis for directly accountable decision-making (Minton 2009; Sennett 2018). The ethical urbanist is required to be simultaneously attentive to social justice, representative accountability and climatic and ecological sustainability. Current primers on place-making illustrate some elements of how this can be achieved (Hamdi 2010; Mostafavi 2017; Chan 2019). As we confront the gradual demise of neoliberalism, ethical cites are also dealing with multiple recurring shocks, whether from pandemics or social, economic and/or climate change. At the same time, critical questions surround what constitutes sustainable economic development as we shift into a new era that may be characterised by prosperity without growth (Jackson 2009) and planned degrowth (Gleeson and Alexander 2018; Nelson and Schneider 2018; Nelson 2018).

Ethical practitioners in inclusive communities

Ethical urbanism is a social construct and is broadly indicative of the right to participate in and be genuinely represented via political processes. It is concerned with relationships between people within communities and grounded in the concept of social recognition where individuals come to recognise their shared commitments and obligations (Brooks 2014). This formulation is problematic, not least because the possibility of shared identities undermines other forms of diversity (race, gender, age, etc.). Ethical urbanism therefore has to incorporate a 'conception that includes both the ideas of unity and diversity' or more specifically unity in diversity based on an understanding that differences enrich human interactions (Hulkko 2005, p. 2). It also implies a relationship beyond membership or 'love of patria, whether the object of attachment is city, country, team, firm or cosmos' (Brown 2015, p. 218). As mentioned in Chapter 3, there needs to be a commitment by all urbanites to a shared ethic negotiated and reconciled within cities in pursuit of the common good. This is complicated, however, in situations where market values are the only values and where people are rebranded as consumers of goods and services, catered for by ever-expanding choices in the marketplace.

The dilemmas of negotiating ethical citizenship are not unique to our times; indeed, they are as ancient as our cities and can be traced back (at least) to the Athenian Oath of the Ephebes, where oath makers swore never to bring disgrace to their city: 'The land-of-my-ancestors [*patris*] I will leave [to the next generation] in a condition that is not diminished but instead greater and better than it had been before' (Nagy 2018). Modern interpretations of the Oath (for instance as shared by the US National League of Cities) stipulate that 'we will transmit this city . . . not less, but greater and more beautiful than it was transmitted to us.' Beauty of course, is in the eye of the beholder, and the ethical city requires concrete measures that show how inequality, sustainability and good governance are addressed. Moreover, ethical people in the city should have a critical understanding of power and engage proactively in deliberation with other conflicting (and often more powerful) interests (Brooks 2014). They are aware and capable of enacting their responsibility to be engaged, to participate and to be represented in pursuit of ethical principles around integrity, equality, deliberation, accountability, stewardship, sustainability, justice and human rights (Brown 2010) – and local authorities have a responsibility to ensure appropriate strategies are in place to develop such capacities (Cuthill and Fien 2005). Ethical urbanism depends upon forms of control by the people derived from trust and transparency mutually reinforced and factored into decision-making. Inevitably, this extends well beyond signing a petition or completing a questionnaire about a proposed freeway. The failures of such faux-democracy tokenism are well known. In the 1960s Sherry Arnstein, employed by the US Government, implemented a range of initiatives designed to enhance public engagement by federal agencies, showing how the most effective forms of public participation involve varying degrees of public control, as expressed in Figure 4.1 in the form of a Ladder of Public Participation. These range from agencies partnering with people to agencies delegating decision-making to residents, to residents deciding how resources are used and priorities managed (Arnstein 1969).

One of the best-known examples of public control is participatory budgeting (introduced in Chapter 3 as an example of preference utilitarianism). A trailblazer example here is that of Mayor Olivia Duto in the city of Porto Alegre in late 1980s Brazil, whose administration implemented such an experiment, which could be placed somewhere across steps 6 to 8 on Arnstein's ladder (de Sousa Santos 1998; Luna 2010). The policy was popular in a city with a culture and history of sovereign resistance and rebellion (Dixon and Sarkees 2015). The Porto Alegre model of participatory budgeting is not defined in law but is based on an agreement – an urban social contract if you will – between residents and the municipal government. It is an annual process conducted across three tiers: from neighbourhood assemblies, to a participatory council of delegates from 17 city regions and finally to a city-level council (Wyngowski 2013). The proposals are subject to mayoral approval, and the role of municipal government is supportive rather than directing. Research by the World Bank found that around one-fifth of the population surveyed had participated in the budgeting process, with lower representation from the poor, youth and most surprisingly middle-high income groups (World Bank 2008). In 2011, Porto Alegre was ranked as the second (after Curitiba) most budgetarily transparent of the 27 provincial capital cities in Brazil (INESC 2011).

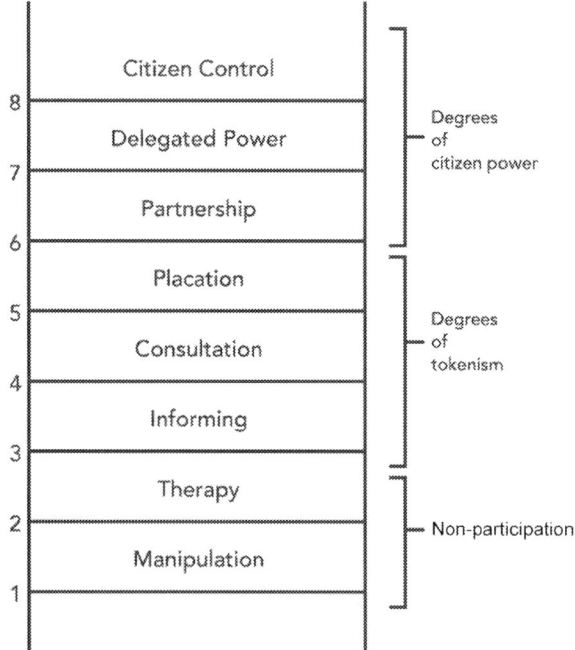

FIGURE 4.1   Arnstein's ladder of participation

Source: Arnstein (1969, p. 217).

Ethical community entrepreneurs

Normative ideas of entrepreneurship are shaped by neoliberal rationalities and expressed through language, ideals, imaginaries and politics. Market rationalities, entrepreneurship and employability are seen as inherently ethical because the role of citizens is to become the best market/neoliberal subjects they can possibly be (Bloom 2017, p. 13). Entrepreneurship has become so pervasive that it even covers an ever-growing gamut of people engaged in social private-for-public-good enterprises. Further, within entrepreneurship two broad groups are evident. The first incorporates entrepreneurial qualities crucial to ethical decision-making (imagination, creativity, novelty, etc.). This can be understood as productive entrepreneurship. The second exemplifies darker traits including obsession with personal financial gain even if this involves violating social norms, corruption and crime. This is best described as unproductive or destructive entrepreneurialism (Baumol 1990; Harris et al. 2009). Countries with high levels of corruption seemingly have lower levels of productive entrepreneurship (Avnimelech et al. 2011; Avnimelech and Zelekha 2015).

Ethical entrepreneurialism has broader implications than the traditions of individuals, start-ups and business exploitation to create goods and services for profit (Zucchella et al. 2018). Amongst these are the implied need for the ethical entrepreneur to practice deontological dimensions of ethics (Chapter 3) by using the ideas of entrepreneurship, such as innovation and marketing, to make the world more sustainable (Bloom 2017). It thus reflects the on-going efforts of individuals, enterprises, communities and cities involved in transformative experiments and practices designed to address social and environmental problems with an awareness of the future implications of their actions and a desire to minimise harm and maximise social and environmental sustainability. This connects with social entrepreneurs and shared value (Porter and Kramer 2011) but extends them deontologically in prioritising people and planet over profit.

People-centred approaches to economic development, such as in community wealth building, aim to establish supportive frameworks for ethical entrepreneurs (examined in Chapter 8). For example, the Centre for Local Economic Strategies (CLES), collaborating with local governments in the UK, employs five strategies to strengthen local economies: (1) encouraging plural ownership of the local economy by small enterprises, community organisations, cooperatives and municipalities; (2) ensuring financial power works for local places by harnessing and recirculating investment (e.g. by using the existing wealth in local authority pension funds) rather than attracting external capital; (3) promoting fair employment and just labour markets including payment of a living wage; (4) facilitating progressive procurement of goods and services to create dense supply chains of local businesses that support local employment, small enterprises, employee-owned businesses, social enterprises, cooperatives and community businesses and (5) supporting socially just use of land and property and community use of public sector land and facilities, thus deepening ownership of local assets so that any financial and social gain benefits those who live in the community (Centre for Local Economic Strategies 2019).

The origins of this approach can be traced back to the Mondragon Corporation, a federation of worker cooperatives founded in the Basque region of Spain in 1956. The Mondragon Corporation, which has a presence in 97 countries, brings together over 80,000 workers in 266 companies and cooperatives – hence functioning as an advanced industrial economy (Bamburg 2017). Governed by a general assembly, Mondragon's initial design was highly democratic and relatively fair with every worker as an owner and entrepreneur, with clearly defined procedures to establish wage differentials and transparent formulae for the distribution of profits. Perhaps the most important point is that Mondragon was designed to be 'organic and evolutionary, not ideological or utopian' (Stikkers 2011). The Mondragon model, though challenging to replicate, supports ethical entrepreneurship characterised by the ability to obtain strength from collective vision and experience, a focus on long-term strategy and where profit is a means to serve members of the collective (Clamp and Alhamis 2010).

Another example, the Democracy Collaborative based in Cleveland, USA, reflects the particularities of community-based economies in US cities, including an emphasis on place, ownership, multipliers, collaboration, inclusion, workforce and systemic dimensions (Kelly and McKinley 2015). The latter point is particularly significant for ethical entrepreneurship since the goal is to develop institutions and ecosystems that support the creation of new economic activity. This includes municipalities supporting community-owned banks to increase lending to small businesses. The basic assumption is that adoption of a community wealth drivers approach, based upon strategies such as anchor institution procurement, financing, enterprise development and retention, land and real estate and so on, has the corollary effect of enhancing traditional approaches that tend to be more focused on the individual such as entrepreneurship programmes, workforce developments and the creation of incubators. What these developments in the UK, Spain and the United States imply is that community-based entrepreneurs are emerging as a third force, in addition to business and government, in shaping local economic development. These new players are the antithesis of mobile capital and include a diverse range of anchor institutions such as universities, local government, health service providers, trade unions, large local businesses and the voluntary sector such as housing associations (Kelly and McKinley 2015; Centre for Local Economic Strategies 2017).

Challenges confronting ethical city shapers

The idea of promoting ethical practice in the shaping of cities has particular currency in the contemporary conditions of urban crises (Moraitis and Rassia 2019). The integrity of our institutions of governance lies at the root of many of these crises. This includes government itself, but also the structures of governance in the world of finance. According to the OECD, public trust in government is at an all-time low (OECD 2013). Compiling data from a range of public opinion surveys (World Gallup Poll, World Values Survey, Eurobarometer), the OECD found in 2012 that only six out of every ten people expressed confidence in their national governments. This data highlights public sentiment four years after the 2008 GFC, the management of which was widely recognised as having overwhelmingly rewarded perpetrators and punished innocents (Petruno 2013). In this context, the Edelman Trust Barometer (ETB) is a reliable source of long-term data. The barometer reveals that post-GFC, levels of trust in government worldwide dropped to as low as 43 percent in 2009, rising to 52 percent in 2011 but then dropping back to 44 percent in 2014 (World Economic Forum 2015). There had been some improvement with the 2020 ETB showing trust in government at 47 percent for the mass population and 59 percent for the informed public (i.e. college educated, significant consumers of media, engaged in public policy).

ETB results consistently indicate that government performs poorly in terms of level of public trust compared to NGOs, business and even the media. Indeed, government is identified as the least competent and most unethical of all four (Edelman 2020). This could explain why Barak Obama described contemporary politics and leadership as 'characterised by corruption, carelessness, self-dealing, disinformation, ignorance and just plain meanness' (Scott 2020). In the

United States, trust in government in 2017 stood at an all-time low of only 18 percent (Pew Research Center 2019). Similarly, an analysis of public sentiment from international surveys covering over four million people in 154 countries between 1995 and 2020 revealed dissatisfaction with governments at record high levels in 2019 (Foa et al. 2020). These findings imply that contemporary levels of democratic dissatisfaction in the UK, Australia, Brazil, Mexico and the United States are one third higher than they were in the 1990s. The researchers concluded that:

> [i]f confidence in democracy has been slipping, then the most likely explanation is that democratically elected governments have not been seen to succeed in addressing some of the major challenges of our era, including economic coordination in the eurozone, the management of refugee flows, and providing a credible response to the threat of global climate change.
> *(Foa et al. 2020, p. 42)*

Aside from questions of what constitutes good democratic governances, anti-elite populism and anti-system politics are significant forces reflecting deep feelings of unfairness about who wins and who loses in contemporary society (Hopkin 2020; Eatwell and Goodwin 2018). Populism (which takes both right- and left-wing forms) represents a trust-deficit in governments everywhere and juxtaposes people against a self-serving and corrupt elite in its various guises. Liberal democracies are hampered by flawed systems of representation and media controlled by dominant private interests. Populist rejection of the status quo can be readily guided in the opposite direction of fair and transparent, representative governance. This was clearly demonstrated by the conspiracy stories that emerged around the US 2016 elections related to election integrity and voter fraud (Norris et al. 2020). Moreover, 'Drain the swamp' might only mean replacing old swamp-dwellers with new – or, to mix metaphors, a matter of swamps and mirrors. Analysis undertaken by political scientists in collaboration with *The Guardian* identified a surge in populist rhetoric from political leaders in 40 countries from 1998 onwards, as expressed in the language and framing used in their speeches (Hawkins *et al.* 2019). The rubric used in analysing these speeches highlights how some populist leaders convey a Manichaean (we are good, other are evil) view of the world portrayed as a struggle between darkness and light. Every issue is ascribed a strong moral dimension of right or wrong, with nothing in between, no nuance allowed and no shades of grey. This is compounded by a situation where nationalist populism is rooted in four Ds – (1) *Distrust* of established political and economic elites and belief that ordinary people no longer have a meaningful voice; (2) *Deprivation* in the sense that large segments of society (we) are being left behind relative to others, and where the liberal elite devotes far more attention to the others, abandoning the 'we'; (3) *Destruction* manifesting in fears that national groups, identities and ways of life are under threat – manifesting in the notion that 'we and our values are under attack' and (4) *De-alignment* occurring when bonds between 'we the people' and their traditional political parties begin to breakdown (Eatwell and Goodwin 2018). Local government in cities across the world are not immune to these conditions. Data indicates very low turnouts (less than 20 percent) in elections in many North American cities (LSE Cities 2014). Some European cities make the 30–50 percent range. Cities with relatively higher turnout for local elections appear to be largely located in Northern Europe (Sweden, Denmark, Belgium) and Latin America (Brazil, Uruguay, Peru). Turnout is certainly no insulator against forces of reaction.

Distrust in politics is also tied into levels of corruption which manifests on an alarming scale and is linked to societal inequality and discrimination based on class, gender, race and sexual orientation compounded by economic disadvantage. The potential loss to the global economy could be of the order of between 2 and 5 percent, implying a figure close to US$2 trillion (Thomson 2017). In 2016, the International Monetary Fund indicated that bribery alone had an annual cost over US$1.5 trillion (IMF 2016). A 2017 Transparency International survey revealed that one in every four people worldwide has had to pay a bribe when accessing public services in the previous 12 months (Transparency International 2017). The survey also found that on an average globally, the police are considered to be the most corrupt, followed by elected representatives, government officials, business executives and local government. Petty corruption is dwarfed by legal, semi-corporate activities, from tax havens to laundering and, simply, not asking questions or turning-a-blind-eye. At the city level, Transparency International labelled London as the number-one home for the fruits of corruption because of its role as a safe haven for corrupt money from across the globe (Shenker 2016). This raises the concern of how much of your city skyline buildings are the result of dumping large amounts of cash surpluses from suspect deals that leave a trail of environmental and human misery. In the face of broken and corrupt political systems, populism plumbs new depths in the manipulation of rights and vulnerabilities of ordinary people (Mijuskovic 2017). There are however more ethical pathways that could emerge tied into renewal of democracy and cosmopolitan conceptions of peoplehood (Wolkenstein 2019). These pathways are borne out of the failure of liberal democratic institutions to address widening inequalities and point to a key tenet of the ethical city – namely, the existence and active maintenance of a shared set of agreed and inclusive rules governing how we treat each other, find common cause and pursue mutual benefits.

However, if we take a wider view of leadership and governance and see them as characterised by shared responsibility for decision-making, such as proposed in ethical cities, then public trust in leadership can soar. This is seen in two studies on local leadership in Australia. The first focused on community leadership and found that the leaders of community organisations, such as service and sporting clubs and various neighbourhood associations, were seen as more trustworthy, more selfless and more concerned for social, economic and environmental well-being than political, religious, trade union and business leaders (Fien and

Wilson 2015). A particularly significant finding in this study related to the capacity of community leaders to choose values-based approaches to 'wicked problems' rather than the coercive and managerialist approaches favoured by other type of leaders (p. 23). The second study focused on local government and found that 42 percent of Australians trusted their local council compared with 31 and 28 percent, respectively, who said they trusted their state and national governments (Essential Report 2018). Recent results from the ETB confirmed this latter finding as being international in that local levels of government in 18 of the 24 countries surveyed were seen as more ethical than national governments – the exceptions were Kenya, Brazil, Indonesia, India, South Korea and Mexico (Edelman 2020, p. 41). The respondents were asked the following question: 'In thinking about why you do or do not trust [INSTITUTION], please specify where you think they fall on the scale between two opposing descriptions.' These opposing descriptions are actually ethical/unethical perceptions as presented in Table 4.5.

Prospects for ethical city shaping

The complex problems facing cities signify the inevitability and urgency of transformative change. These changes could occur in response to shocks such as pandemics or climate change impacts and would be exacerbated by a general failure to implement preparedness measures or to adapt. Likewise, changes could also be associated with a plethora of ethical city experiments. The latter is preferable in minimising overall harm to humans and ecosystems, while maximising the transition towards increased levels of sustainability. Fortunately, numerous experiments and innovations designed to respond to the changing nature of employment, patterns of economic activity and to build local economic resilience are underway in cities. Some of these experiments are community-based,

TABLE 4.5 Measuring ethics in institutions

Dimension	Ethical Perception		Unethical Perception
Purpose-driven	Highly effective agent of positive change	_____	Completely ineffective agent of positive change
Honest	Honest and fair	_____	Corrupt and biased
Vision	Has a vision for the future that I believe in	_____	Does not have a vision for the future that I believe in
Fairness	Serves the interests of everyone equally and fairly	_____	Serves the interests of only certain groups of people

Source: Edelman (2020, p. 74).

requiring that conflicting interests collaborate closely. Indeed, cities are arguably key sites for the emergence of new forms of circular- and social-economy-based enterprises that support diversity and solution focused approaches to problems of inequality, accountability and climate change (Mouraviev and Kakabadse 2016; Mouraviev and Kakabadse 2019).

In some cases, these transformations are taking shape in cities through sharing economies, which are socio-economic ecosystems built around the collective use of human, physical and intellectual resources (McLaren and Agyeman 2015; Nelson 2018). It is a term used to describe an extensive range of people-centric, local economic activities including swapping, exchanging, collectively purchasing, collaboratively consuming, shared ownership, cooperatives, co-creation, recycling, upcycling, redistribution, trading used goods, renting, borrowing, lending and so on. These new economic patterns of collaborative consumption provide access to idle or underutilised assets for ridesharing, co-working or urban agriculture and strengthen the social fabric through deliberative decision-making and active citizenship (Botsman and Rogers 2011). While in some instances sharing economy platforms undermine decent working conditions, casualising and dehumanising workers, there are alternative models for governance and accountability in sharing cities that promote fairness and inclusivity and also deliver urban resilience, economic interdependence and enhanced social cooperation (Sharp 2016).

City governments are starting to extend their role as facilitators to attract and retain private sector investment to new roles in encouraging social enterprises and entrepreneurship. The City of Oslo, for example, procures more than US$3 billion worth of goods and services every year and as part of this ensures ethical contract requirements in procurement supply chains are met. The City has a track record of developing ambitious social and environmental projects and plans to slash carbon emissions by 95 percent by 2030, divest its $9 billion pension fund from fossil fuels, remove car traffic from parts of the city centre, power all public transport from renewables and spend $1 billion to upgrade and expand the city's tram system (King 2016; Pashley 2015; Berglund 2016). All of these initiatives send a strong and clear ethical message to the residents of Oslo – these measures are the right thing to do. Deeper transitions are implied as cities seek to generate new forms of local economic development and new industries, without sacrificing the integrity of existing communities to poverty, homelessness, drugs or other forms of socio-economic breakdown. In the United States, movements such as the Metropolitan Revolution and the Democracy Collaborative involve local leaders growing the job market and making their communities more prosperous (Katz and Bradley 2013; Kelly and McKinley 2015). This requires physical resources and political will to ensure that community plays a central role in city economic development. In Chapter 8 we will elaborate further and share more ethical city examples that include the redirection of public expenditure to promote local jobs, the introduction of local currencies, the emergence of circular economies and a range of universal basic income experiments.

Open governance, accountability, transparency and anti-corruption provide a solid foundation for ethical conduct and, from the consequentialist perspective (Chapter 3) outcomes.

While the course of transitions cannot be known, neither can the outcomes of a concerted ethical city shaping, nor whether it is likely to manifest at sufficient speed to avoid a climate catastrophe. However, what is more known is the unpredictable and disturbing forms of contemporary populist political engagement. Transparency, accountability, empowerment and principled action to address unfairness, inequality and environmental destruction are all key areas for the renewal of local political processes. Practical actions include citizen assemblies and urban social contracts based on accountability and transparency on the part of local political leaders and government. The challenge is to reinvent civic leadership, planning, economic development and public engagement in the multicultural, cosmopolitan, pluralistic, ethical city. To bring this to reality, there is a significant onus upon key actors or shapers of the ethical city, including city and multilevel governance actors/leaders, urban planners/designers/developers, residents and entrepreneurial businesses. Inevitably, this points to a need for rapid institutional reform designed carefully to avoid unintended or contradictory consequences for inequality, representation and sustainability.

References

Arnstein, S. (1969) A ladder of citizen participation, *Journal of the American Planning Association 35* (4): 216–224.

Avnimelech, G. and Zelekha, Y. (2015) The impact of corruption on entrepreneurship, in I. Management Association (ed.) *Business Law and Ethics: Concepts, Methodologies, Tools, and Applications.* Hershey: IGI Global.

Avnimelech, G., Zelekha, Y. and Sarabi, E. (2011) *The Effect of Corruption on Entrepreneurship,* paper presented at the DRUID 2011 on Innovation Strategy and Structure: Organisations, Institutions, Systems and Regions at the Copenhagen Business School, Denmark, June 15–17. [Online]. Available: https://conference.druid.dk/acc_papers/1 944qlhkqrqpnsq5gmkf4yvuy4m2.pdf [Last accessed: 10 May 2020].

Bamburg, J. (2017) Mondragon through a critical lens: Ten lessons from a visit to the Basque cooperative confederation, *Medium Magazine.* [Online]. Available: https://medium.com/fifty-by-fifty/mondragon-through-a-critical-lens-b29de8c6049 [Last accessed: 6 February 2020].

Barber, B.R. (2013) *If Mayors Ruled the World: Dysfunctional Nations, Rising Cities.* New Haven and London: Yale University Press.

Barcelon en Comú (2014) Governing by obeying - Code of political ethics. [Online]. Available: https://barcelonaencomu.cat/sites/default/files/pdf/codi-etic-eng.pdf [Last accessed: 6 February 2020].

Barrett, B.F.D., Horne, R. and Fien, J. (2016) The ethical city: A rationale for an urgent new urban agenda, *Sustainability 8* (1197): 1–14. https://doi.org/10.3390/su8111197.

Baumol, W.J. (1990) Entrepreneurship: Productive, unproductive, and destructive, *The Journal of Political Economy 98* (5): 893–921. https://doi.org/10.1086/261712.

Berglund, N. (2016) Oslo plans major tram line expansion, views and news from Norway. [Online]. Available: www.newsinenglish.no/2016/09/12/oslo-plans-major-tram-expansion/ [Last accessed: 7 April 2020].

Blanco, I., Salazar, Y. and Bianchi, I. (2019) Urban governance and political change under a radical left government: The case of Barcelona, *Journal of Urban Affairs 42* (1): 18–38.

Bloom, P. (2017) *The Ethics of Neoliberalism: The Business of Making Capitalism Moral.* Routledge Studies in Business Ethics, London and New York: Routledge.

Botsman, R. and Rogers, R. (2011) *What's Mine Is Yours: How Collaborative Consumption Is Changing the Way We Live.* New York: Collins.

Brooks, T. (ed.) (2014) *Ethical Citizenship: British Idealism and the Politics of Recognition.* Basingstoke and New York: Palgrave Macmillan.

Brown, M.T. (2010) *Civilizing the Economy: A New Economics of Provision.* Cambridge: Cambridge University Press.

Brown, W. (2015) *Undoing the Demos: Neoliberalism's Stealth Revolution.* New York: Zone Books.

Calvino, I. (1972) *Invisible Cities.* London: Vintage Books, published in 1997.

Centre for Local Economic Strategies (2017) *Community Wealth Building through Anchor Institutions.* Manchester: CLES. [Online]. Available: https://cles.org.uk/wp-content/uploads/2017/02/Community-Wealth-Building-through-Anchor-Institutions_01_02_17.pdf [Last accessed: 11 May 2020].

Centre for Local Economic Strategies (2019) *Community Wealth Building 2019: Theory, Practice and Next Steps.* Manchester: CLES. [Online]. Available: https://cles.org.uk/wp-content/uploads/2019/09/CWB2019FINAL-web.pdf [Last accessed: 11 May 2020].

Chan, J.K.H. (2019) *Urban Ethics in the Anthropocene.* Basingstoke and New York: Palgrave Macmillan.

Clamp, C.A. and Alhamis, I. (2010) Social entrepreneurship in the Mondragon cooperative corporation and the challenges of successful replication, *The Journal of Entrepreneurship 19* (2): 149–177. https://doi.org/10.1177/097135571001900204.

Cuthill, M. and Fien, J. (2005) Capacity building: Facilitating citizen participation in local governance, *Australian Journal of Public Administration 64* (4): 63–80.

De Sousa Santo, B. (1998) Participatory budgeting in Porto Alegre: Towards a redistributive democracy, politics and society, Stoneham. [Online]. Available: www.ssc.wisc.edu/~wright/santosweb.html [Last accessed: 7 April 2020].

Dixon, J.S. and Sarkees, M.R. (2015) *A Guide to Intra-State Wars: An Examination of Civil, Regional and Intercommunal Wars, 1816–2014.* Los Angeles, London, New Delhi, Singapore and Washington: Sage Publishing.

Eatwell, R. and Goodwin, M. (2018) *National Populism: The Revolt against Liberal Democracy.* London: Pelikan Books and Penguin Random House.

Edelman (2020) Edelman trust barometer: Global report 2020. [Online]. Available: www.edelman.com/sites/g/files/aatuss191/files/2020-01/2020%20Edelman%20Trust%20Barometer%20Global%20Report.pdf [Last accessed: 15 April 2020].

Essential Report (2018) *Trust in Institutions.* [Online]. Available: https://essentialvision.com.au/trust-in-institutions-11 [Last accessed: 15 April 2020].

Fainstein, S.S. (2010) *The Just City.* Ithaca: Cornell University Press.

Fien, J. and Wilson, S. (2015) *The Swinburne Leadership Survey: Index of Leadership for the Greater Good.* Melbourne: Swinburne University.

Foa, R.S., Klassen, A., Slade, M., Rand, A. and Williams, R. (2020) *The Global Satisfaction with Democracy Report 2020.* Cambridge: Centre for the Future of Democracy.

Gleeson, B. and Alexander, S. (2018) *Degrowth in the Suburbs: A Radical Urban Imaginary.* Basingstoke and New York: Palgrave Macmillan.

Goodchild, B. (1990) Planning and the modern/postmodern debate, *The Town Planning Review 61* (2): 119–137. https://doi.org/10.3828/tpr.61.2.q5863289k1353533.

Hamdi, N. (2010) *The Placemaker's Guide to Building Community.* Washington: Earthscan.

Harris, J.D., Sapienza, H.J. and Bowie, N.E. (2009) Ethics and entrepreneurship, *Journal of Business Venturing 24*: 407–418. https://doi.org/10.1016/j.jbusvent.2009.06.001.

Hawkins, K.A., Aguilar, R., Castanho Silva, B., Jenne, E.K., Kocijan, B. and Rovira Kaltwasser, C. (2019) *Measuring Populist Discourse: The Global Populism Database*, paper presented at the 2019 EPSA Annual Conference in Belfast, UK, June 20–22. [Online]. Available: https://populism.byu.edu/App_Data/Publications/Global%20 Populism%20Database%20Paper.pdf [Last accessed: 10 May 2020].

Healey, P. (2000) Planning in relational space and time: Responding to new urban realities, in Bridge, G. and Watson, S. (eds.) *A Companion to the City.* Oxford: Blackwell Publishing.

Hopkin, J. (2020) *Anti-System Politics: The Crisis of Market Liberalism in Rich Democracies.* Oxford: Oxford University Press.

Hulkko, A. (2005) *Democracy and Ethical Citizenship, Politics of Participation: Focus on the Third Sector, Civil Engagement and Dimensions of Citizenship*, Helsinki, August. [Online]. Available: www.helsinki.fi/project/eva/pop/pop_Hulkko.pdf [Last accessed: 6 February 2020].

IMF (2016) *Corruption: Costs and Mitigation Strategies*, Staff Team from the Fiscal Affairs Department and the Legal Department. [Online]. Available: www.imf.org/external/ pubs/ft/sdn/2016/sdn1605.pdf [Last accessed: 7 April 2020].

INESC (2011) *Budget Transparency in Brazilian Capitals*, Instituto de Estudos Socioeconômicos, Brasilia. [Online]. Available: www.internationalbudget.org/wp-content/uploads/ Budget-Transparency-in-Brazilian-Capitals.pdf [Last accessed: 7 April 2020].

Jackson, T. (2009) *Prosperity without Growth: Foundations for the Economy of Tomorrow.* London and New York: Routledge.

Jacobs, J. (1961) *The Death and Life of Great American Cities.* New York: Random House.

Katz, B. and Bradley, J. (2013) *The Metropolitan Revolution: How Cities and Metros Are Fixing Our Broken Politics and Fragile Economy.* Washington: Brookings Institution Press.

Kelly, M. and McKinley, S. (2015) Cities building community wealth, *Democracy Collaborative.* [Online]. Available: https://democracycollaborative.org/learn/publication/ cities-building-community-wealth [Last accessed: 23 April 2020].

Kidder, P. (2008) The urbanist ethics of Jane Jacobs, *Ethics, Place & and Environment 11* (3): 253–266. https://doi.org/10.1080/13668790802559668.

King, E. (2016) Oslo votes to slash emissions 95% by 2020, *Climate Home News.* [Online]. Available: www.climatechangenews.com/2016/06/23/oslo-votes-to-slash-emissions- 95-by-2030/ [Last accessed: 7 April 2020].

Langlois, L. (2011) *The Anatomy of Ethical Leadership: To Lead Our Organizations in a Conscientious and Authentic Manner.* Edmonton: Athabasca University Press.

LSE Cities (2014) *How Cities Are Governed: Building a Global Databased for Current Models of Urban Governance.* [Online]. Available: https://urbangovernance.net/en/ [Last accessed: 7 April 2020].

Luna, M. (2010) Participatory budgeting: Sharing power over public resources, *Shareable.* [Online]. Available: www.shareable.net/participatory-budgeting-sharing-power- over-public-resources/ [Last accessed: 18 March 2020].

McLaren, D. and Agyeman, J. (2015) *Sharing Cities: A Case for Truly Smart and Sustainable Cities.* Cambridge, MA: MIT Press.

Mijuskovic, O.Z. (2017) The ethical implications of populism in political practice, *Global Ethics Network: Rethinking International Relations*, Carnegie Council for Ethics in International Affairs. [Online]. Available: www.globalethicsnetwork.org/profiles/blogs/the-ethical-implications-of-populism-in-political-practice [Last accessed: 4 February 2020].

Minton, A. (2009) *Ground Control: Fear and Happiness in the Twenty-First-Century City.* London: Penguin.

Moraitis, K. and Rassia, S.T. (2019) *Urban Ethics under Conditions of Crisis: Politics, Architecture, Landscape Sustainability and Multidisciplinary Engineering.* Singapore: World Scientific Publishing Company.

Mostafavi, M. (ed.) (2017) *Ethics of the Urban: The City and the Spaces of the Political.* Zurich: Lars Muller Publishers.

Mouraviev, N. and Kakabadse, N.K. (2016) Conceptualising cosmopolitanism and entrepreneurship through the lens of the three-dimensional theory of power, *Society and Business Review 11* (3): 242–256.

Mouraviev, N. and Kakabadse, N.K. (eds.) (2019) *Entrepreneurship and Global Cities: Diversity, Opportunity and Cosmopolitanism.* New York and London: Routledge.

Nagy, G. (2018) The oath of the ephebes as a symbol of democracy: And of environmentalism, *Classical Inquiries.* [Online]. Available: https://classical-inquiries.chs.harvard.edu/the-oath-of-the-ephebes-as-a-symbol-of-democracy-and-of-environmentalism/ [Last accessed: 7 April 2020].

Nelson, A. (2018) *Small is Necessary: Shared Living on a Shared Planet.* London: Pluto Press.

Nelson, A. and Schneider, F. (2018) *Housing for Degrowth: Principles, Models, Challenges and Opportunities.* Abingdon and New York: Routledge.

Nemeth, J. and Hollander, J. (2010) Security zones and New York's city's shrinking public space, *International Journal of Urban and Regional Research 34* (1): 20–34. https://doi.org/10.1111/j.1468-2427.2009.00899.x.

Norris, P., Garnett, H.A. and Grömping, M. (2020) The paranoid style of American elections: Explaining perceptions of electoral integrity in an age of populism, *Journal of Elections, Public Opinion and Parties 30* (1): 105–125. https://doi.org/10.1080/17457289.2019.1593181.

OECD (2013) *Government at a Glance 2013.* Paris: OECD Publishing.

O'Toole, J. (1996) *Leading Change: The Argument for Values-Based Leadership.* New York: Ballantine Books.

Pashley, A. (2015) Norway's oil-rich capital first to divest from fossil fuels, *Climate Home News.* [Online]. Available: www.climatechangenews.com/2015/10/19/norways-oil-rich-capital-first-to-divest-from-fossil-fuels/ [Last accessed: 7 April 2020].

Petruno, T. (2013) 5 years after financial crash, many losers: And some big winners, *Los Angeles Times.* [Online]. Available: www.latimes.com/business/la-fi-crisis-winners-losers-20130915-story.html [Last accessed: 4 February 2020].

Pew Research Center (2019) *Public Trust in Government 1958–2019.* [Online]. Available: www.people-press.org/2019/04/11/public-trust-in-government-1958-2019/ [Last accessed: 7 April 2020].

Porter, M.E. and Kramer, M.R. (2011) The big idea: Creating shared value, *Harvard Business Review.* [Online]. Available: https://hbr.org/2011/01/the-big-idea-creating-shared-value [Last accessed: 10 May 2020].

Price, T.L. (2008) *Leadership Ethics: Introduction.* Cambridge: Cambridge University Press.

Rapoport, E., Acuto, M. and Grcheva, L. (2019) *Leading Cities: A Global Review of City Leadership.* London: UCL Press.

RMIT University (2016) *The Rotorua Model: Interview with Mayor Steve Chadwick at the Ethical Cities Urban Thinkers Campus, Melbourne.* [Online]. Available: https://youtu.be/diBH3ufJx10 [Last accessed: 25 August 2020].

Rode, P. (2019) *Climate Emergency and Cities: An Urban-Led Mobilization?* LSE Cities. [Online]. Available: https://lsecities.net/wp-content/uploads/2019/10/Rode-P-2019-Climate-Emergency-and-Cities-An-urban-led-mobilisation.pdf [Last accessed: 10 May 2020].

Sager, T. (2011) Neo-liberal urban planning policies: A literature survey 1990–2010, *Progress in Planning* 76: 147–199. https://doi.org/10.1016/j.progress.2011.09.001.

Sassen, S. (2016) How Jane Jacobs changed the way we look at cities, *The Guardian.* [Online]. Available: www.theguardian.com/cities/2016/may/04/jane-jacobs-100th-birthday-saskia-sassen [Last accessed: 6 February 2020].

Scott, E. (2020) Barack Obama's endorsement of Joe Biden, annotated, *The Washington Post.* [Online]. Available: www.washingtonpost.com/politics/2020/04/14/transcript-obama-biden-endorsement/ [Last accessed: 16 April 2020].

Sennett, R. (2018) *Ethics for the City: Building and Dwelling.* New York: Allen Lane, an imprint of Penguin Books.

Sharp, D. (2016) Sharing cities: Why ownership, governance and the commons matter more than ever, *Shareable.* [Online]. Available: www.shareable.net/blog/sharing-cities-why-ownership-governance-and-the-commons-matter-more-than-ever [Last accessed: 18 April 2016].

Shenker, J. (2016) Which are the most corrupt cities in the world? *The Guardian.* [Online]. Available: www.theguardian.com/cities/2016/jun/21/which-most-corrupt-cities-in-world [Last accessed: 4 October 2020].

Singer, P. (2011) *Practical Ethics*, 3rd Edition. Cambridge: Cambridge University Press.

Stiglitz, J.E. (2019) After neoliberalism, *Project Syndicate.* [Online]. Available: www.project-syndicate.org/commentary/after-neoliberalism-progressive-capitalism-by-joseph-e-stiglitz-2019-05 [Last accessed: 6 February 2020].

Stikkers, K.W. (2011) Dewey, economic democracy, and the Mondragon cooperatives, *European Journal of Pragmatism and American Philosophy III-2.* [Online]. Available: http://journals.openedition.org/ejpap/833 [Last accessed: 6 February 2020.

Thomson, S. (2017) We waste $2 trillion a year on corruption: Here are four ways to spend that money, *World Economic Forum.* [Online]. Available: www.weforum.org/agenda/2017/01/we-waste-2-trillion-a-year-on-corruption-here-are-four-better-ways-to-spend-that-money [Last accessed: 7 April 2020].

Thompson, W.N. and Leidlein, J.E. (2008) *Ethics in City Hall: Discussion and Analysis for Public Administration.* Sudbury, Ontario and London: Jones & Bartlett Publishers.

Transparency International (2017) *Global Corruption Barometer: Citizens' Voices from Around the World.* [Online]. Available: https://www.transparency.org/news/feature/global_corruption_barometer_citizens_voices_from_around_the_world [Last accessed: 7 April 2020].

Wechsler, R. (2013) *Local Government Ethics Programs: A Resource for Ethics Commission Members, Local Officials, Attorneys, Journalists, and Students, and a Manual for Ethics Reform.* North Haven, CT: City Ethics Inc.

Wolkenstein, F. (2019) Populism, liberal democracy and the ethics of peoplehood, *European Journal of Political Theory* 18 (3): 330–348. https://doi.org/10.1177/1474885116677901.

World Bank (2008) *Brazil: Toward a More Inclusive and Effective Participatory Budget in Porto Alegre. 2008, Volume 1. Main Report.* [Online] Available: https://openknowledge.worldbank.org/handle/10986/8042 [Last accessed: 30 August 2016).

World Economic Forum (2015) *Outlook on the Global Agenda 2015.* [Online]. Available: http://reports.weforum.org/outlook-global-agenda-2015/global-leadership-and-governance/global-leadership-index/ [Last accessed: 15 April 2020].

Wright, I.L. (2013) *Are We All Neoliberals Now? Urban Planning in a Neoliberal Era,* proceedings ISOCARP Congress 2013 on Frontiers of Planning, Brisbane Australia, 1–4 October, pp. 1384–1407. [Online]. Available: www.isocarp.net/Data/case_studies/2412.pdf [Last accessed: 10 May 2020]. Updated version available: www.cbp.com.au/services/planning,-infrastructure-and-environment/planning-government-infrastructure-environment-g.

Wyngowski, S. (2013) Local participation in Brazil: Porto Alegre's model for 21st century local government, *Gnovis Journal.* [Online]. Available: www.gnovisjournal.org/2013/12/11/local-participation-in-brazil-porto-alegres-model-for-21st-century-local-government/ [Last accessed: 10 May 2020].

Zucchella, A., Hagen, B. and Serapio, M.G. (2018) *International Entrepreneurship.* Basingstoke and New York: Edward Elgar Publishing.

5

ASSESSMENT OF THE ETHICAL CITY

> No one, wise Kublai, knows better than you that the city must never be confused with the words that describe it. And yet between the one and the other there is a connection.
>
> Italo Calvino, *Invisible Cities*, 1972, p. 53

Localisation of the Sustainable Development Goals

At face value the Sustainable Development Goals (SDGs), also known as the 2030 Agenda, provide a common framework that can be readily adopted by cities in pursuit of prosperity, sustainability and good governance. The goals (see Figure 5.1) follow decades of such initiatives from which an extensive range of indicators, standards, guidelines and principles were developed. The question is, what difference can they make when incorporated into the tools of ethical city governance? As city governments consider how best to prioritise and integrate the SDGs within their existing practices and how best to monitor progress made in pursuit of these goals, questions are arising about the localisation and contextualisation of the SDGs and short-comings in the operationalisation of the goals. Several metaphors are useful in reflecting on these questions, including 'Shifting Deckchairs', 'Reinventing the Wheel', 'Top-down Approach', 'Measuring Everything' and 'Bearing the Costs'.

> *Shifting Deckchairs:* The SDGs are ostensibly a reworked (or repurposed) version of Agenda 21 and the Millennium Development Goals. Agenda 21 was the main outcome document from the 1992 Rio Earth Summit (UN Conference on Environment and Development) – but it was a non-binding action plan. The Millennium Development Goals were

FIGURE 5.1   Sustainable development goals

Source: UN (2020).

adopted at the UN Millennium Summit in 2000 and included eight goals on poverty reduction, hunger eradication, universal primary education, gender equality, reducing child mortality, improving maternal health, combatting HIV/AIDS and other diseases, ensuring environmental sustainability and developing global partnerships. Under the voluntary Agenda 21 implementation scheme, each member state of the United Nations was encouraged to submit national reports. Following the 2012 UN Conference on Sustainable Development (Rio+20) a three-year long dialogue took place on what the next steps in international cooperation towards sustainable development should be. At the same time, a High-Level Political Forum (HLPF) on Sustainable Development, which meets under the auspices of the UN General Assembly and the UN Economic and Social Council, was formed. These deliberations led to the UN General Assembly approving the 2030 Agenda and endorsing the SDGs. A Division for Sustainable Development Goals was established as the designated agency responsible for supporting national voluntary reviews (NVRs) on the implementation of the SDGs. The participation rate in NVRs is extensive but as of 2020, contains notable exceptions including the United States, Iran and Cuba.

Reinventing the Wheel: City governments have been encouraged to prepare local voluntary reviews (LVRs) under the SDG process. These have a lineage back to the 1992 Rio Earth Summit and the establishment of the Local Agenda 21 (LA21) process in cities and towns across the globe, under the auspices of Local Governments for Sustainability (ICLEI). By 2012, over 6400 local governments in 113 countries worldwide were engaged in LA21, with the vast majority (80 percent) in Europe (Rok and Kuhn 2012). Spanning an era when the development of local consultative processes were being popularised, numerous challenges in trying to facilitate and sustain community engagement were documented (Barrett and Usui 2002). Fast forward and LVRs share many characteristics with LA21 in terms of raising visibility of sustainability concerns, ensuring local government relevance and distinguishing their contribution as the local manifestation of the 2030 Agenda. A new banner with the title Local 2030 has emerged in the form of a multi-stakeholder network and platform that supports on-the-ground delivery of the SDGs through local governments and their networks, international organisations, national governments, civil society organisations, academic institutions and the private sector.

Top-Down Approach: The 17 SDGs, and the extensive associated targets, have been described as complex, often un-implementable and contradictory (Liverman 2018). It is unclear who is responsible for what and who is accountable. This failing could result, for example, in the SDGs having limited impact on reducing inequalities (Pogge and Sengupta 2016). The SDGs also project a form of cockpit-ism – deluded thinking that suggests 'top-down steering by governments and international organizations alone can address global problems' (Hajer et al. 2015, p. 1652). A critical concern here is that the SDGs fail to directly confront existing structural problems, power imbalances and ideological biases underpinning our international development system. Given that the SDGs were not conceived of as a means to challenge systemic oppression, alter power relations and undermine patterns of exploitation, they may do little to advance social justice and foster ethical development (Sultana 2018). A search of the 2015 UN resolution on the SDGs reveals only a single mention of 'democracy' and 'human rights'. Critically important terms such as 'civil liberties', 'free expression', 'press freedom', 'independent judiciary', 'separation of powers' and 'free and fair elections' are absent (Smith and Gladstein 2018). 'Corruption' is mentioned only twice, including a vague and unmeasurable target under Goal 16 to 'substantially reduce corruption and bribery in all their forms' (UN 2015). The SDGs, after all, were developed by UN processes and therefore are embedded in prevailing structures of political and economic power (Block 2018).

Measuring Everything: Our world is awash with sustainability indicators that have emerged over the past three decades since the 1992 Rio Earth Summit. We have indicators for almost every topic imaginable, developed by UN agencies, the World Bank, OECD, EU, international NGOs, national governments and their agencies, local governments and their associations and the private sector. While these indicators are important for policy goal-setting and performance monitoring, their proliferation poses a significant challenge for practitioners at the local level (Shen et al. 2011). This has led to attempts at rationalisation, with researchers at the National University of Singapore reducing some 1,100 known urban indicators down to core group of 158 in 13 categories (Malone-Lee 2016). In 2014, ISO 37120 was released, focusing on a set of 100 indicators under 17 themes for city services and quality of life. Early cities in this process included Rotterdam, Johannesburg, Buenos Aires and Guadalajara, with certification managed by the World Council on City Data (Ng 2016). Nevertheless, generally such indicators suffer from both technical and conceptual inconsistencies (Huang et al. 2015). Many do not incorporate a triple bottom line approach, fail to capture external/leakage effects beyond city boundaries and lack the ability to assess cities in both developing and developed countries using a common approach (Mori and Christodoulou 2012). Most importantly, urban sustainability indicators collectively create a shrill white noise that obscures the instrumentalities of urban unsustainability.

Bearing the Costs: The Sustainable Development Solutions Network (SDSN) estimates it would cost in the region of US$1 billion per annum to enable 77 of the world's lower-income countries to put in place statistical systems capable of supporting and measuring the SDGs (SDSN 2015). Even at the local level, research on the data required to report in five cities (Bangalore, Cape Town, Gothenburg, Manchester and Kisumu) found that all faced significant problems (Simon *et al.* 2015). Many localities will be challenged in the assessment realm. While reliable estimates for the costs of monitoring the SDGs progress across the cities of the world are elusive, inevitably such activities are more affordable for the well-resourced global cities which are generally already participating in one or more of the many city-network initiatives (C40, UCLG, ICLEI, etc). Secondary and smaller cities, of which there are thousands across the globe, will find it much harder to finance SDG assessment and reporting.

Voluntary local reviews

This brings us to the consideration of how the SDGs are being localised and monitored. Efforts to localise the SDGs in the first cycle of the 2030 Agenda (2015–2019) have been extensively documented (SDSN 2016) and are coordinated under the framework of the Global Taskforce of Local and Regional Governments. This Taskforce is facilitated by United Cities and Local Governments (UCLG) – a network founded in 2004 and comprising over 240,000 towns, cities and regions and over 175 local government associations. Reporting to the HLPF, the Taskforce asserts that VLRs should gain greater support internationally and nationally (UCLG 2019). The first VLRs were pioneered in 2018 by New York City and a cluster of Japanese cities – Kitakyushu, Toyama and Shimokawa – with the latter closely following the methodology set out in the UN's Handbook for the

Preparation of Voluntary National Reviews (UN DESA 2019). Subsequently, there has been a gradual uptake of VLRs as shown in Figure 5.2.

Analysis of these nascent VLRs reveals a number of interesting characteristics. First, there is considerable variation in methodology, format and content. There are instances where VLRs (1) lack local political support (mayoral endorsement); (2) fail to align with other city strategies; (3) do not incorporate a public awareness raising function around the SDGs; (4) neglect to elaborate a partnership approach and (5) are unclear on the financing/budgeting arrangements (Pipa and Bouchet 2020). Second, four key institutional models have been adopted to support VLR development: (1) One Key Office/Team – preferably in the Mayor's office coordinating the initiative; (2) Hub and spoke – one coordinator with outreach to relevant offices within the local government; (3) Interagency – collaboration between different units within the local authority led by a steering committee or working group and (4) partnership between a city office and external organisation (e.g. a university or think tank). These organisational models seem relatively basic when compared with past Local Agenda 21 engagement models where the emphasis was on extensive stakeholder dialogues and the opening up of previously closed policy formulation processes (Buckingham-Hatfield and Percy 1999; Barrett and Usui 2002). Third, only a small selection of VLRs actually addresses all 17 SDGs, with many tending to focus on a small subset normally associated with thematic issues covered by the annual HLPF (Pipa and Bouchet 2020).

Bristol in the UK is one of the few examples of a city with a VLR that attempts to localise all 17 goals. Its experience is particularly insightful, with Bristol City Council and the Cabot Institute for the Environment at Bristol University producing a handbook to support other local governments interested in engaging in the process (Fox and Macleod 2019a). The handbook points to 12 main reasons to undertake a VLR exercise:

(1) Facilitate work across silos in pursuit of a holistic agenda;
(2) Foster data-led policy proposals;
(3) Ensure a leave-no-one-behind approach focused on helping the most vulnerable in the local community;
(4) Highlight strengths and weaknesses in local government performance;
(5) Create a shared language around issues facing the city;
(6) Enhance multi-sectoral partnerships within the city;
(7) Build stakeholder engagement;
(8) Emphasise transparent accountability;
(9) Facilitate monitoring of progress;
(10) Demonstrate city leadership;
(11) Provide opportunities for shared learning amongst cities and
(12) Exemplify global citizenship

The Bristol VLR (2019) is enhanced by its collaboration with the Bristol Green Capital Partnership, a community interest company founded in 2007 that brings

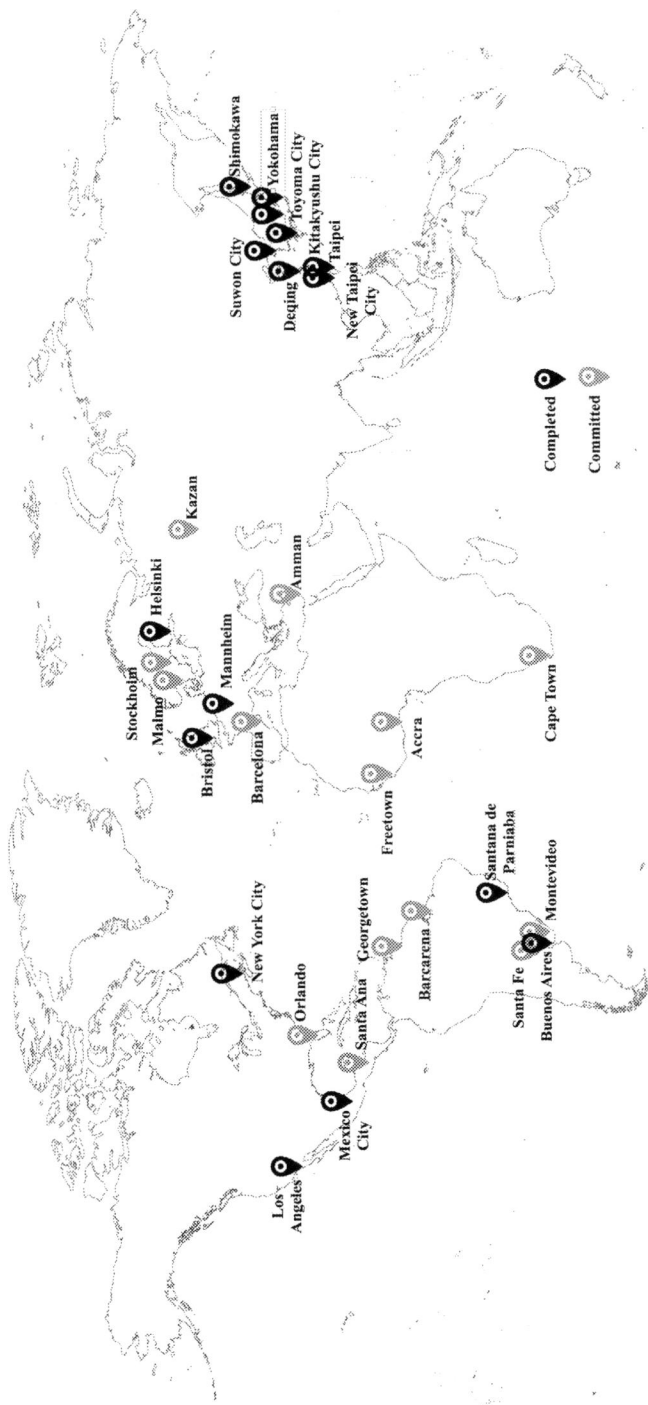

FIGURE 5.2 VLRs completed or underway in the first cycle of the 2030 Agenda

Source: Redrawn from Pipa and Bouchet (2020, p. 2) using Copyright Free Vector Maps.com.

together over 900 member organisations. The challenge in the Bristol case has been to cross reference local policies embodied in the Bristol One City Plan (published in January 2019 – discussed further in Chapter 8) with the SDGs and to facilitate cross-sectoral implementation (Deininger et al. 2019). However, the Bristol VLR process raises a number of concerns around jurisdictional boundaries, what should be measured, data deficits, the appropriateness of indicators and challenges of data disaggregation (Fox and Macleod 2019b; Fox and Macleod 2019c). Thus, one report makes a more sober assessment on three localisation challenges related to policy coherence, public engagement and inconsistencies within the goals (Carden-Noad 2017). This report noted the immense challenge involved in aligning national, regional and local policies with the SDGs, the failure of the goals to gain traction with the general public and the fact that the SDGs leave unaddressed the contradiction between promoting sustained economic growth and promoting environmental protection. Rather the report's authors suggest that the VLR focus should be on targeting areas of worst deprivation and inequality with policies to promote community economic growth.

How do we assess progress towards the ethical city?

Notwithstanding the challenges and shortcomings of goals and their assessment in cities, there is a simple utility at the heart of such ideas; they provide a common language and a sense of progress. The all-important questions are around what is the object and purpose of the common language, who is involved and engaged, and how and what is to be measured by whom? The challenges associated with VLR suggest that perhaps we need to devote more attention to these questions in order to guide ethical urban development. In 2015, the authors developed an ethical city diagnostic tool referred to as the Ethical City Scan (ECS). Like the later VLR, the aim of the ECS was to assist with city reporting requirements, in this case for city signatories to the UN Global Compact Cities Programme. The ECS was designed as an online diagnostic tool for cities to self-analyse progress across the ten Global Compact principles (see Table 5.1). It originally utilised 157 issue indicators broken down into 22 categories under three core themes of city development, sustainability and governance (see Annex 1). As shown in Table 5.2, this compares with a total of 232 indicators for the SDGs, but is considerably higher than other indicator frameworks. Implicit within ECS is the notion that measuring progress towards the ethical city should be given due weight in resourcing, decision-making and policy priority setting across the city. Also, importantly, a focus was on processes of co-production being a collaborative exercise involving extensive co-design with community groups and city partners, plus various departments across the local authority.

The City Scan methodology requires the collective local ranking of each issue on a Likert scale. In other words, there is a deliberate qualitative component. The resultant assessment, however, is not meant to form the basis for comparison across cities or league tables of performance. Rather, the assessment represents a baseline

TABLE 5.1 Ten principles of the UN Global Compact

Human Rights

Principle 1: Support and respect the protection of internationally proclaimed human rights; and

Principle 2: Not be complicit in human rights abuses.

Labour

Principle 3: Uphold freedom of association and the effective recognition of the right to collective bargaining;

Principle 4: The elimination of all forms of forced and compulsory labour;

Principle 5: The effective abolition of child labour;

Principle 6: The elimination of discrimination in respect of employment and occupation.

Environment

Principle 7: Support a precautionary approach to environmental challenges;

Principle 8: Undertake initiatives to promote greater environmental responsibility; and

Principle 9: Encourage the development and diffusion of environmentally friendly technologies.

Anti-Corruption

Principle 10: Work against corruption in all its forms, including extortion and bribery.

Source: UN Global Compact (2016, p. 6).

TABLE 5.2 Indicator framework for the Ethical City Scan, SDGs and others

Indicator Framework	Number of Indicators	Number of Cities
Ethical City Scan	157	19
SDGs	232	N/A
World Council on City Data ISO37120	100	79+
United Smart Cities Smart Sustainable Cities	90	50+
IESCE Cities in Motion Index	79	180
SDSN USA Cities Index 79	44	100
Arcadis Sustainable Cities	32	100
Indicators for Sustainability	32	11
UN Habitat City Prosperity Index	25	400
Urban Ecosystem Europe	25	32

Source: Fox and Macleod (2019b, p. 15).

for future within-city comparisons and a reflection of local discussions about how the city is, can and should perform on particular indicators. The respondents can also identify city strengths – areas where the local government is perform-ing well. In addition, respondents are encouraged to submit detailed qualitative comments on each issue, with supporting documentation, plans and policies. This more extensive documentation assists in evaluating the impact of local policies, plans and practices. Implementation of ECS (version 1) in the 2015/16 period elicited responses from 19 city signatories, collectively representing a population of 16 million people and including cities ranging in size from as small as 21,000 population (more of a town) up to close to four million (Figure 5.3).

A ranking of the top ten issues based on the original pilot ECS results is presented in Table 5.3. These are compared to rankings of the same issues made by participants in the 2016/2017 ethical cities MOOC (number enrolled totalled 4,059 with 10 percent completing the survey). For comparative purposes, we include results from a 2014 survey of 202 cities undertaken by researchers at the LSE Cities Programme on the main challenges facing cities in the next ten years (Rapoport et al. 2019). While the research designs and the survey respondents varied, there are notable commonalities in the issues identified. First, housing affordability and poverty ranked highly, as do concerns around economic vitality, unemployment and various dimensions of urban mobility. This said, the fact that ECS predominantly covered cities in the Global North could explain the overwhelming concern for housing affordability and this may be different for other cities (e.g. see the Leading Cities Survey results in Table 5.3). Second, when local government respondents complete the survey, we postulate they are less likely to overtly criticise their city, especially in relation to matters of corruption or malpractice. There are exceptions, however, as illustrated by the results from the Barcelona ECS (see Figure 5.4) where the respondent identified a number of critical issues in the area of city governance, including lack of accountability of leaders and bureaucrats as well as concerns around transparency of various processes within local government. Third, as with any assessment tool with an intentionally subjective orientation, results are only as good as the process, and local political dynamics can result in certain issues being viewed as particularly sensitive and lead to biased or null returns.

However, the strength of such qualitative tools is that they rest upon inclusive processes, and their value lies in ethical practice and engagement. The data and benchmark criteria help to anchor each indicator assessment for future reference and iterations. However, they are not intended to deliver equal weightings attached to specific issues over time, and they do not provide summable or comparable results across indicators (criteria and scales vary across people, social group and cultures). They also exhibit spacetime contingency (criteria are somewhat implicit and value laden) and do not indulge in the pretence of 'objective' reality (scores are accepted as being based on judgement and local knowledge albeit with the use of benchmark information). The utility of such assessments resides in their processual function and transparency, rather than their compliance with a utopian objective reality that many city assessment tools aspire to (Veenhoven 2002).

There is merit in evaluating urban social policies through a combination of both objective/quantitative and subjective/qualitative approaches given that one alone cannot provide sufficient information (Santos and Martins 2007; Veenhoven 2002). In this context, the ECS was designed to allow cities and regions to support their subjective claims with objective and subjective indicators. One example to elucidate this point can be found in a study on urban quality of life

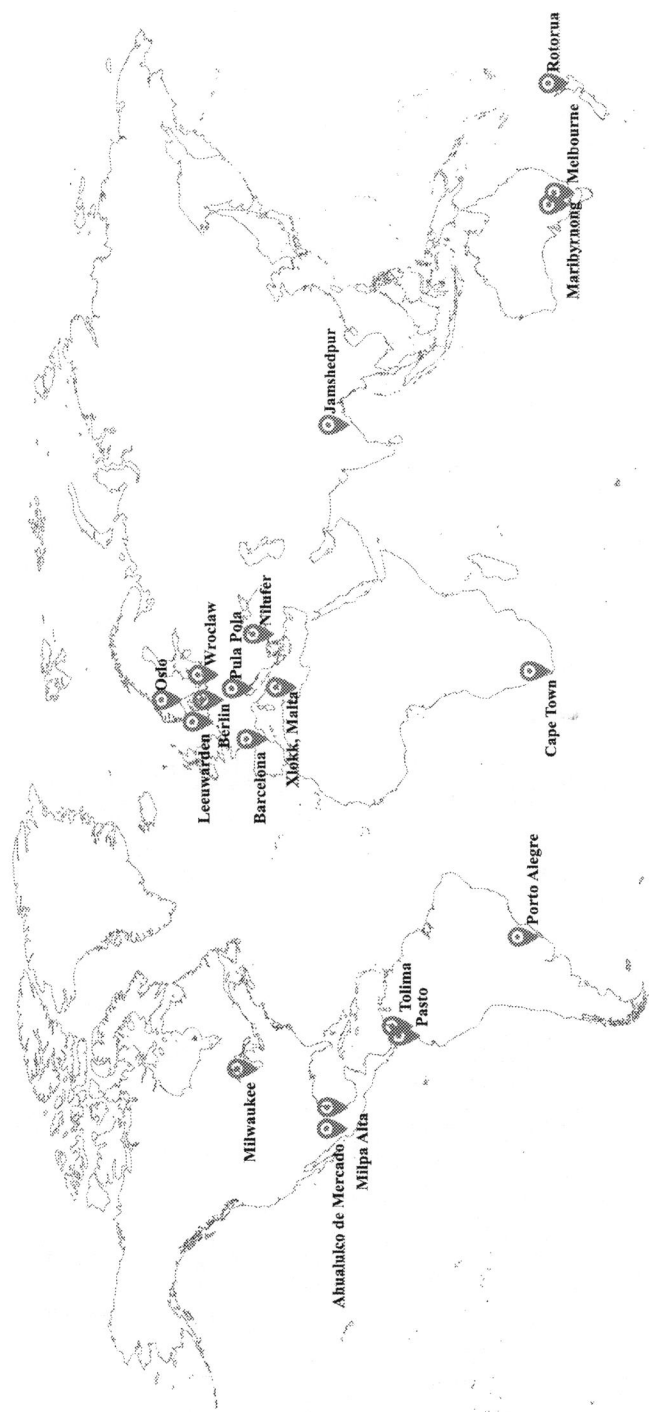

FIGURE 5.3   Cities surveyed in the 2015/2016 Ethical City Scan

Source: Global Compact Cities Programme. Copyright Free Vector Maps.com.

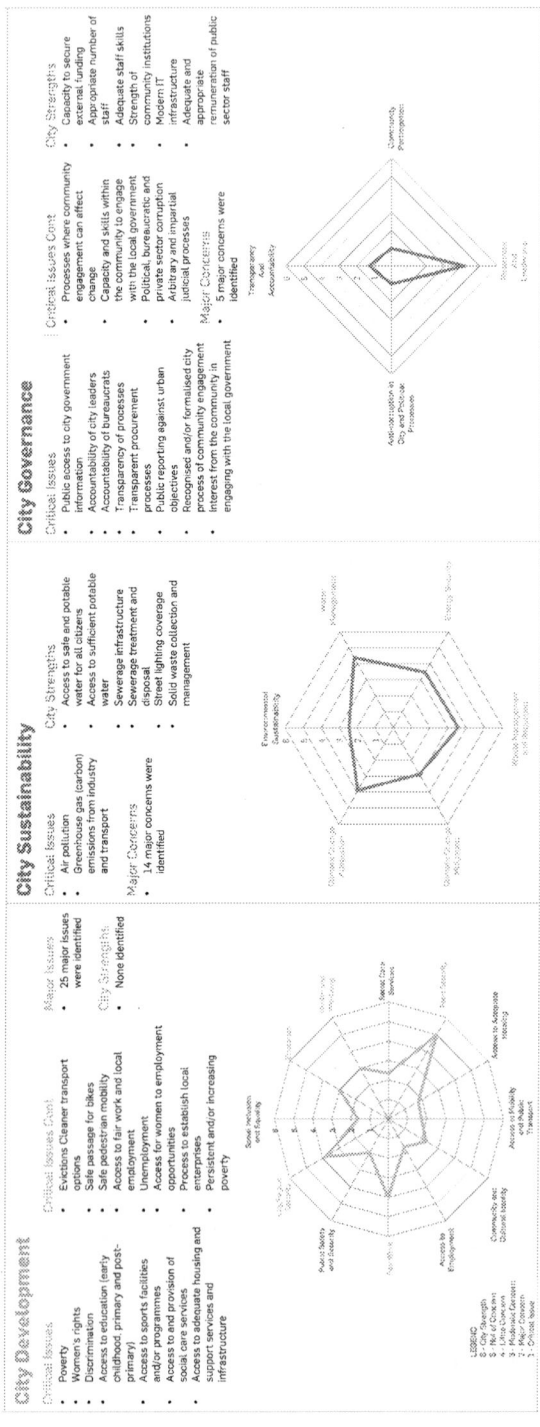

FIGURE 5.4 Summary of the Barcelona ECS results 2015

Source: Global Compact Cities Programme (2015).

TABLE 5.3 Top ten challenges facing cities

Ethical City Scan 2015	MOOC Respondents 2016/2017	Leading Cities Survey 2014
Lack of affordable housing	Lack of affordable housing	Mobility and urban connectivity (45%)
Safe passage for bikes	Poverty	Demographic changes and social equality (43%)
Poverty	Public transport connectivity	Urban planning (40%)
GHG emissions from transport	Unemployment	Poverty and living costs (28%)
Unemployment	Dependency on fossil fuels	Economic vitality (23%)
Community understanding of climate change	Safe passage for bikes	Lack of affordable housing (23%)
Safe pedestrian mobility	Cleaner transport options	Environmental sustainability (22%)
Access to local employment	Clean energy alternatives	Effective governance (18%)
Clean energy alternatives	Access to fair work	Urban safety and security (12%)
Cleaner transport options	Safe pedestrian mobility	Regional and international issues (11%)

Source: Derived from Rapoport et al. (2019).

in Porto Alegre, Brazil that compared a crime rate indicator to residents' perceptions of urban insecurity and crime (Santos and Martins 2007). While the crime rate data demonstrated that the city was comparatively safe, residents considered insecurity as the most negative aspect of urban living. This example reveals why it is important for surveys to include an opportunity for respondents to provide extensive qualitative information related to the issues under examination.

A weakness of all indicators is their bounded and constraining nature. The authors asked participants in the Ethical Cities MOOC in 2016 and 2017 to identify additional issues (beyond the 157 issues covered in the ECS). As a result, a further 564 issues were identified, analysed and grouped (see summary word cloud shown in Figure 5.5). One of the main concerns raised related to community cohesion (identified by 26 participants). Key concerns that emerged from this exercise included corruption (24 counts), inequality (23 counts), urban sprawl (19), weak planning system (17), traffic congestion (15), air pollution (14) and homelessness (14). Education (21 counts) is another issue raised by the survey respondents covering a broad range of topics including need, cost and quality.

What to measure and how?

The VLR experience to date points to questions over what to measure and around who decides and with what consequence? Across this contested and problematic

FIGURE 5.5 Word cloud of additional issues identified by MOOC participants

Source: Global Compact Cities Programme (Produced with www.jasondavies.com/wordcloud/).

terrain, an ethical city must inevitably navigate a path that is both principled and outcome oriented. The act of measurement is itself purposive, subjective and laden with politics and, as a consequence, ethical city assessment tasks should be linked to actions aligned to structural changes that address root causes rather than symptoms. Let us take public safety from economic crime (including juvenile crime) and violence (including domestic violence) as an example. One potential benefit of the surveillance society is its potential to enhance the visibility of crime and violence across the city (despite the inevitable loss of liberties and rights). However, cameras do not make a safe city, as so-called 'safe' designed precincts shift crime and violence as efficiently as gentrification shifts the urban poor out to make way for more wealthy classes. In cities with CCTV saturated streetscapes we can imagine how internal domestic spaces that are (so far) under less surveillance, become even more concentrated sites for crime and violence than they are already. The irony is one of streets replacing homes as our places of safe sanctuary (notwithstanding that for many, home was never safe). Nevertheless, in the hope of marginal gains in public safety, Crime Prevention Through Environmental Design (CPTED) has become a core plank of some planning systems (Knapp 2013). In the UK, for example, this involves police certification of buildings and car parks promoted through national legislation such as the Sustainable and Secure Buildings Act 2004. This approach completely sidesteps the poverty-related causes of the vast majority of crimes. In Portugal, as another example, urban security concerns are finding form in strategic integrated planning of the Lisbon Metropolitan area, or put more simply, the aim is to enforce the basic right of urban residents not to be the victim of crime (Tulumello 2016). The question for the ethical city is how to shift the causes of crime and violence in tandem with innovation in urban spaces. This emphasis on crime, violence, security and other concerns that impact on city liveability and fragility is discussed further in Chapter 6.

A central challenge in assessing progress towards the ethical city is that it represents a combination of processes, outcomes, structures, agency, alignments and intent. This means assessment cannot be reduced to an off-the-shelf solution; rather it suggests a positioning for how assessment might be done differently from identifying normative goals, endpoints and outputs. Since the ethical city is principled, assessment must elaborate those principles. Since the ethical city is locally experienced and located, the aims, emphases and expected progress should be agreed locally for each city (and arguably in each neighbourhood or locality across the city). The ethical city is ultimately about ensuring the public is in the driving seat, not some remote global consultants or local technocrats. Adopting normative indicators unquestioningly is unlikely to lead to meaningful progress; much of the work on assessment is in analysing what really needs to be measured. Nevertheless, if we accept that there are limits and planetary boundaries that we must live within, then this suggests we correspondingly need to choose local environmental sustainability indicators that reflect this reality, including carbon, energy and water footprints (Hajer et al. 2015), as well as considerations related to socio-economic distributional justice, including social indicators such as the Gini coefficient of income and the poverty ratio (Mori and Yamashita 2014).

Carbon intensity and income inequality are logical key meta-indicators, although care is required in how these are aggregated across populations and over time. Furthermore, governance accountability measures are increasingly available to identify obviously corrupt practices. For example, Transparency International promotes an initiative to strengthen local government integrity (McDevitt 2014). Across all these domains, we stress the need to consider four key tenets in ethical city measurement:

- The ethical city is a principled concept. There is neither a single blueprint nor a useful league table of ethical cities. Measurement is necessary only as a longitudinal, progress-based exercise to test what is working or not working in each given city. To state the obvious: every city is different and will create its own ethical future.
- Both objective quantifiable measures and more qualitative measures involving engagement, judgement and discussion have a critical role in measurement. They should allow associated actions to engage with structural causes rather than symptoms.
- Qualitative measures should be accompanied by broad engagement across civil society, technical experts, policymakers and other city stakeholders in assessing progress and priorities for the ethical city.
- Residents need to co-plan and co-direct the measurement of progress towards the ethical city. All too often the approaches that have been developed remain

in the hands of politicians and technocrats, whereas ethical local governments should actually be at the forefront of new, innovative modes of community engagement, so that residents can make sound decisions with the best available advice.

Need for ethical engagement around the SDGs

While there are numerous tools available for measuring progress on the SDGs, they address only part of the problem of city assessment at best; at worst they represent a top-down technocratic approach that allows the problems of cities to be outsourced and the solutions to be avoided. To be successful, the SDGs and related VLRs should enhance engagement between cities, communities and residents. Ethical conversations and reflections are important conduits for the materialisation of the ethical leadership of ideas, since ethical city practices shape, and are shaped by, ethical communities. Standard tools of reflexive governance point to engaged processes that create niche experiments and share the results, revising plans and actions to constantly re-adjust priorities and actions (Voss et al. 2006). A long history of such activity is punctuated by aforementioned initiatives such as LA21 and a variety of city networks promoting engaged assessment (C40, ICLEI, Global Compact of Mayors, Rockefeller 100 Resilient Cities, UN Global Compact Cities Programme, etc).

The outcome of such approaches is unpredictable, and the future course of cities is potentially un-steerable. However, ethical city development is processual in charting a course of openness and inclusion in the businesses of governing, building cross-sectoral coalitions, monitoring progress and reflecting upon it and being open to changing tactics and experimenting along the way. Cities can advance, test and publish successful civic engagement strategies, sharing knowledge with progressive private, public and NGO/civil society organisations and incorporating innovative techniques for engagement with communities. Cities can work with partners to evaluate progress, assessment methodologies and tools; to develop recommendations and guidelines and to explore the implications of ethical city policies, programmes and models. Cities can engage with local researchers to co-produce a range of scholarship and reflective conversations about ethical city models and future implications for wider academic, policy and public discussion (Polk 2015). Furthermore, they can advance ideas about transitions to the ethical city, which include notions of inclusiveness and respect for diversity as societal goals.

Ethical cities construct open spaces for deliberative governance, transparency and accountability while weeding out corruption, conflicts of interest and the abuse of power. They encourage measures aimed at maintaining public trust and fairness in local government action. In turn, they use their financial spending power and leadership to promote ethical practices within the city boundaries so that the local economy, including social and community enterprise, can flourish. Local governments recognise the civic responsibilities of individual residents and facilitate their commitment to civic duties. They seek to overtly build the capacity of urban professionals to conduct and manage processes for ethical city development. In an era of unprecedented urbanisation, there is a monumental opportunity to influence the fate of cities through well-designed, sustainable and co-managed infrastructure, housing and urban places. The city is in a position to advance ethical design, planning and developer communities through education, participation, good governance and accountability, often by simply using tools and resources that already exist (RICS 2015). The ethical cities agenda advances a principles-based and collaborative approach to urban development for government actors, private sector, civil society, individuals and communities. Making space for each other, literally and figuratively, is a central concept for the ethical city. We are concerned that cities focusing on maintaining business as usual (pursuing short term, partisan or private economic or political objectives) are at risk of escalating social disaffection, poverty and corruption. They risk spiralling social breakdown, rapidly declining liveability and an unsustainable future. Adoption of the SDGs offers one step on a possible pathway forward, with the caveat that concerns for social justice and ethical engagement within cities are put at the core, rather than as a sidebar.

References

Barrett, B.F.D. and Usui, M. (2002) Local agenda 21 in Japan: Transforming local environmental governance, *Local Environment* 7 (1): 49–67. https://doi.org/10.1080/13549830220115411.

Block, T. (2018) *Pitfalls of the SDGs*. [Online]. Available: www.cdo.ugent.be/blog/pitfalls-sdgs [Last accessed: 25 February 2020].

Buckingham-Hatfield, S. and Percy, S. (eds.) (1999) *Constructing Local Environmental Agendas: People, Places and Participation*. London and New York: Routledge.

Calvino, I. (1972) *Invisible Cities*. London: Vintage Books, published in 1997.

Carden-Noad, S., Macleod, A., Skidmore, T. and Turner, E. (2017) *Bristol and the UN Sustainable Development Goals: An Assessment of Strategic Alignment, Data and Delivery Gaps in Bristol*, University of Bristol, Bristol Green Capital Partnership and Bristol City Council. [Online]. Available: https://bristolgreencapital.org/wp-content/uploads/2017/07/UoB_SDGs-Bristol-report_final_20-Jul-2017.pdf [Last accessed: 10 May 2020].

Deininger, N., Lu, Y., Griess, J. and Santamaria, R. (2019) *Cities Taking the Lead on the Sustainable Development Goals: A Voluntary Local Review Handbook for Cities*, Heinz College, Carnegie Mellon University. [Online]. Available: www.brookings.edu/wp-content/uploads/2019/07/VLR_Handbook_7.7.19.pdf [Last accessed: 10 May 2020].

Fox, S. and Macleod, A. (2019a) *Voluntary Local Reviews: A Handbook for UK Cities: Building on the Bristol Experience*, Cabot Institute for the Environment, University of Bristol and Bristol City Council. [Online]. Available: www.bristol.ac.uk/media-library/sites/cabot-institute-2018/documents/uk-cities-voluntary-local-review-handbook.pdf [Last accessed: 10 May 2020].

Fox, S. and Macleod, A. (2019b) *Bristol and the SDGs: A Voluntary Review of Progress 2019*, Cabot Institute for the Environment, University of Bristol and Bristol City Council. [Online]. Available: www.local2030.org/library/680/Bristol-and-the-SDGs-a-Voluntary-Local-Review-of-progress-2019.pdf [Last accessed: 10 May 2020].

Fox, S. and Macleod, A. (2019c) *Aligning Bristol's One City Plan with the SDGs*, Cabot Institute for the Environment, University of Bristol. [Online]. Available: https://static1.squarespace.com/static/5b4f63e14eddec374f416232/t/5cb648ab71c10b68661d288f/1555450031449/LDASI-England-Bristol_April19.pdf [Last accessed: 10 May 2020].

Global Compact Cities Programme (2015) Barcelona City Council, Spain: City Scan Summary Report, GCCP, Melbourne.

Hajer, M., Nilsson, M., Raworth, K., Bakker, P., Berkhout, F., de Boer, Y., Rockstrom, J., Ludwig, K. and Kok, M. (2015) Beyond cockpit-ism: Four insights to enhance the transformative potential of the Sustainable Development Goals, *Sustainability 7*: 1651–1660. https://doi.org/10.3390/su7021651.

Huang, L., Wu, J. and Yan, L. (2015) Defining and measuring urban sustainability: A review of indicators, *Landscape Ecology 30*: 1175–1193. https://doi.org/10.1007/s10980-015-0208-2.

Knapp, J. (2013) Safety and urban design: The role of CPTED in the design process, *Safer Communities 12* (4): 176–184. https://doi.org/10.1108/SC-07-2013-0015.

Liverman, D.M. (2018) Geographic perspectives on development goals: Constructive engagements and critical perspectives on the MDGs and the SDGs, *Dialogues in Human Geography 8* (2): 168–185. https://doi.org/10.1177/2043820618780787.

Malone-Lee, L.C. (2016) *An Adaptive Sustainability Assessment Tool for Benchmarking Cities and Evaluating Urban Development Programs*, presentation to the Global Platform for Sustainable Cities (GPSC) Working Group on Indicators for Sustainable Cities & Geospatial Tools, March 7–8, Singapore.

McDevitt, A. (2014) *Local Integrity Assessment System Toolkit*. Berlin: Transparency International. [Online]. Available: www.transparency.org/whatwedo/publication/local_integrity_system_assessment_toolkit [Last accessed: 10 May 2020].

Mori, K. and Christodoulou, A. (2012) Review of sustainability indices and indicators: Towards a new City Sustainability Index (CSI). *Environmental Impact Assessment Review 32* (1): 94–106. https://doi.org/10.1016/j.eiar.2011.06.001.

Mori, K. and Yamashita, T. (2014) Methodological framework of sustainability assessment in City Sustainability Index (CSI): A concept of constraint and maximisation indicators, *Habitat International 45*, 10–14.

Ng, H. (2016) *The World Council on City Data and ISO 37120: An International Platform for Standardized City Indicators*, presentation to the Global Platform for Sustainable Cities (GPSC) Working Group on Indicators for Sustainable Cities & Geospatial Tools, March 7–8, Singapore.

Pipa, A.F. and Bouchet, M. (2020) *Next Generation Urban Planning: Enabling Sustainable Development at the Local Level through Voluntary Local Reviews (VLRs)*. Washington: Brookings Institution. [Online]. Available: www.brookings.edu/research/next-generation-urban-planning-enabling-sustainable-development-at-the-local-level-through-voluntary-local-reviews-vlrs/ [Last accessed: 10 May 2020].

Pogge, T. and Sengupta, M. (2016) Assessing the sustainable development goals from a human rights perspective, *Journal of International and Comparative Social Policy 32* (2): 83–97. https://doi.org/10.1080/21699763.2016.1198268.

Polk, M. (ed.) (2015) *Co-Producing Knowledge for Sustainable Cities: Joining Forces for Change in Cities in South East Africa, Europe and South East Asia*. London and New York: Routledge.

Rapoport, E., Acuto, M. and Grcheva, L. (2019) *Leading Cities: A Global Review of City Leadership*. London: UCL Press.

RICS (2015) *Advancing Responsible Business Practices in Land, Construction and Real Estate Use and Investment*, Royal Institute of Chartered Surveyors/UN Global Compact. [Online]. Available: www.unglobalcompact.org/docs/issues_doc/RICS/GC_RICS_Resource.pdf [Last accessed: 10 May 2020].

Rok, A. and Kuhn, S. (2012) *Local Sustainability 2012: Taking Stock and Moving Forward: Global Review*. Bonn: ICLEI–Local Governments for Sustainability.

Santos, L.D. and Martins, I. (2007) Monitoring urban quality of life: The Porto experience, *Social Indicators Research 80* (2): 411–425. https://doi.org/10.1007/s11205-006-0002-2.

Shen, L.Y., Ochoa, J.J., Shah, M.N. and Zhang, X. (2011) The application of urban sustainability indicators: A comparison between various practices, *Habitat International* 35: 17–29. https://doi.org/10.1016/j.habitatint.2010.03.006.

Simon, D., Arfvidsson, H., Anand, G., Bazaz, A., Fenna, G., Foster, K., . . . Wright, C. (2015) Developing and testing the urban Sustainable Development Goal's targets and indicators: A five-city study, *Environment and Urbanization 28* (1): 49–63. https://doi.org/10.1177/0956247815619865.

Smith, J. and Gladstein, A. (2018) *How the UN's Sustainable Development Goals Undermine Democracy*. [Online]. Available: https://qz.com/africa/1299149/how-the-uns-sustainable-development-goals-undermine-democracy/ [Last accessed: 25 February 2020].

Sultana, F. (2018) An(other) geographical critique of development and SDGs, *Dialogues in Human Geography 8* (2): 186–190. https://doi.org/10.1177/2043820618780788.

Sustainable Development Solutions Network (2015) *Data for Development: A Needs Assessment for SDG Monitoring and Statistical Capacity Development*. New York and Paris: SDSN. [Online]. Available: https://sustainabledevelopment.un.org/content/documents/2017Data-for-Development-Full-Report.pdf [Last accessed: 10 May 2020].

Sustainable Development Solutions Network (2016) *Getting Started with the SDGs in Cities: A Guide for Stakeholders*. New York and Paris: SDSN. [Online]. Available: https://irp-cdn.multiscreensite.com/be6d1d56/files/uploaded/9.1.8.-Cities-SDG-Guide.pdf [Last accessed: 10 May 2020].

Tulumello, S. (2016) Toward a critical understanding of urban security within the institutional practice of urban planning: The case of the Lisbon Metropolitan area, *Journal of Planning Education and Research 37* (4): 397–410. https://doi.org/10.1177/0739456X16664786.

UCLG (2019) *Towards the Localization of the SDGs, Local and Regional Governments' Report to the 2019 HLPF: 3rd Report*. Barcelona: Global Taskforce of Local and Regional Governments. [Online]. Available: www.gold.uclg.org/sites/default/files/Localization 2019_EN.pdf [Last accessed: 10 May 2020].

UN (2015) *Transforming Our World: The 2030 Agenda for Sustainable Development, Resolution by the General Assembly on 25 September 2016,* United Nations, New York. [Online]. Available: https://www.un.org/en/development/desa/population/migration/general assembly/docs/globalcompact/A_RES_70_1_E.pdf [Last accessed: 10 May 2020].

UN (2020) Sustainable Development Goals - Guidelines for the Use of the SDG Logo including the Colour Wheel, and 17 Icons. UN Department of Global Communications, New York. [Online]. Available: https://www.un.org/sustainabledevelopment/wp-content/uploads/2019/01/SDG_Guidelines_AUG_2019_Final.pdf [Last accessed: 10 May 2020].

UN DESA (2019) *Handbook for the Preparation of Voluntary National Reviews: The 2019 Edition*, Department of Economic and Social Affairs, UN, New York. [Online]. Available: https://sustainabledevelopment.un.org/content/documents/20872VNR_hanbook_2019_Edition_v2.pdf [Last accessed: 10 May 2020].

UN Global Compact (2016) *White Paper: The UN Global Compact Ten Principles and the Sustainable Development Goals: Connecting, Crucially,* UN Global Compact, New York. [Online]. Available: https://d306pr3pise04h.cloudfront.net/docs/about_the_gc%2FWhite_Paper_Principles_SDGs.pdf [Last accessed: 10 May 2020].

Veenhoven, R. (2002) Why social policy needs subjective indicators, *Social Indicators Research 58* (1–3): 33–46. https://doi.org/10.1023/A:1015723614574.

Voss, J.-P., Bauknecht, D. and Kemp, R. (2006) *Reflexive Governance for Sustainable Development*. Cheltenham: Edward Elgar.

6

COMPETITIVE, LIVEABLE AND FRAGILE CITIES

> The desire of my Marco Polo is to find hidden reasons which bring men to live in cities: reasons which remain valid over and above any crisis.
>
> Excerpt from a lecture delivered by Italo Calvino at Columbia University in March 1983 (Calvino 2004, p. 181)

Situationism, structure and agency

An Aristotelian idea holds that individuals possess virtues embodying cognitive states and that these determine how we respond to situations (Upton 2009). Several psychological experiments, however, have shown how behaviours can be mediated by our circumstances, including the situation of others around us. The Stanford prison experiment undertaken by the American psychologist, Philip Zimbardo, is a particularly oft-quoted example, where students took on the roles of guards and prisoners. Clearly 'easily unobtrusive situational factors can tap our susceptibilities to obedience, conformity, irresponsibility, cruelty, or indifference to others' welfare' (Badhwar 2009). By implication, therefore, we can anticipate the potential for fundamentally different patterns of behaviour to emerge in today's competitive, liveable cities should the circumstances permit. Moreover, at a structural level, the rules, institutions, practices and routinised incumbency of 'how things are' constrains and channels our actions (Giddens 1984). As such, in this chapter, we turn our attention to the divergent ambitions of cities and, thus, how they shape the behaviours, practices and possibilities for ethical cities. Our basic argument is that city structures, situations and orientations influence how people behave towards each other, how they view themselves and future possibilities and how the city as a diverse community is collectively organised and experienced. Our starting point is to outline four distinct ontological orientations, each based upon contrasting assumptions about what cities ought to be. We acknowledge that more orientations could be advanced; the purpose here is not to propose an exhaustive taxonomy but to illustrate the inherent diversity.

The first is what we will call the competitive city orientation, which we also refer to as Alpha Cities. This orientation is a product of global economic competition in which cities represent the triumph of markets, of competitive advantage and of their relative positioning in attracting talent, capital and other resources. This outward facing view emphasises league tables of economic advantage and imbues a city dogma where urban growth (economically and in terms of population) is an unquestioned good and where the era of urbanisation is celebrated simply for these qualities alone. Needless to say, this orientation de-emphasises internally lived experience, such as inequality, poverty, democratic failings and resource distributional imbalances. There is an implicit assumption that such considerations can be taken care of once the city successfully becomes globally competitive.

The second orientation is what we will call the liveable city orientation where it is assumed that cities are for people and that the quality of their lives is the paramount consideration. Here the focus is on outcomes such as, health, mobility and access to services that afford quality of life and equal opportunities. This offers a more nuanced and essentially inward view of cities that, in turn, requires a more qualitative approach to assessing progress. Most importantly, whilst in the competitive city the implication is that cities are pitched up against each other, in this more collaborative view, there is no such implication.

The third is what can be understood as the fragile city orientation – a relatively new notion applied to thousands of smaller cities across the globe struggling to deal with the negative side effects of rapid economic and population change, as well as the interaction of poverty, violence and vulnerability to natural disasters. Urban fragility can result in (or be the result of) a breakdown in the social contract between the governing authorities and the public. Such cities tend to be located at the bottom of the many indices and league tables of cities (including those that compare liveability as a competitive proposition).

The fourth orientation is that of the fightback cities. These fearless, rebellious and/or ethical cities are engaged actively in both the political critique of the competitive, capitalist project and its shortcomings and in the realisation of alternatives, on the ground, through practising the principles of ethical cities. This orientation was introduced in Chapter 2 and will be further addressed in Chapter 8.

There are numerous economic and geopolitical explanations for the degrees of city competitiveness, liveability and fragility. However, the essential point being made here is that orientation both emphasises and de-emphasises priorities and contains assumptions about what matters most and what takes precedence. In this chapter, we discuss the above orientations and the importance of ethics in all cities (not just the most powerful and globally connected) in order to uncover practices that potentially destabilise urban development across the globe.

Orientation 1: globally competitive alpha cities

The current material and economic realities of cities variously advantage or disadvantage them in the competitive city orientation. Leading urban theorists such as Peter Hall, John Friedmann and Saskia Sassen have shown how the global economy has become structured around a network of 40–50 mega-cities (over ten million population) including New York, London, Paris, Shanghai and Jakarta. The next tier comprises around 600 cities with populations over four million and functioning as competitive dynamos for the global economy (Dobbs et al. 2011). Then there are thousands of smaller cities, many struggling to attract capital and therefore experiencing fundamental difficulties coping with the negative side effects of population growth and failing infrastructures (Muggah 2014a).

City rankings have been popularised through concepts such as the World City System (Friedman 1986), Global City (Hall 1966; Sassen 1991) and Networked City (Castells 2000). The Globalisation and World Cities (GaWC) ranking, for example, provides data on 707 cities (in 2018) and examines their interconnections. This ranking uses a locational analysis of headquarters and branches of transnational corporations to determine the connectivity of each city. The most connected are designated as Alpha++, and include London and New York in the 2018 GaWC ranking. Those with fewest connections are described as Gamma minus (somewhere like Curitiba in Brazil or Penang in Malaysia, for example). This has an Aldous Huxley Brave New World ring to it (Huxley 1932). More importantly, the ranking contains an inevitable bias in that global hierarchies of transnational corporations closely mirror historical and colonial patterns of economic dominance. Indeed, there is a rather telling 2010 GaWC map highlighting a dense concentration of global cities in Europe, North America and to a lesser extent Asia: the rest of the world is, to put it rather bluntly, almost irrelevant in terms of world city connectivity as shown in Figure 6.1.

Contemporary globally connected cities are inevitably a reflection of global capital flows and contests. As such they share key characteristics (Friedmann 1986):

(1) They are integrated into the world economy and, in turn, this impacts on their structure;

(2) Transnational corporations strategically exploit opportunities offered by global cities. For example, Singapore is used as a financial operations centre for major corporations due to its strategic taxation benefits;

(3) They can control influential global functions. For example, New York has evolved into a centre for information, media, entertainment and culture (e.g. CNN in the Times Warner Centre, Fox News on the Avenues of the Americas, and the Huffington Post on Broadway);

(4) They are major sites for the concentration and accumulation of international capital. This is perhaps best exemplified by the race to have the most powerful stock exchanges in London, New York, Tokyo, Shanghai, Hong Kong, Frankfurt and so on;

▲ Map of Global Cities 2010. The map clearly shows areas of the world rather dense and others almost irrelevant in terms of world city connectivity

FIGURE 6.1   Mapping connectedness of global cities

Source: Silvio Carta and Marta Gonzalez (2010): Accessed online on 18 May 2020: www.lboro.ac.uk/gawc/visual/globalcities2010.html).

(5) They are destination points for large numbers of domestic and/or international migrants. In many global cities, such as Sydney, London and New York, 'migrants represent over a third of the population and, in some cities such as Brussels and Dubai, migrants account for more than half of the population' (International Organization for Migration 2015, pp. 38–39);

(6) They reinforce spatial and class polarisation. This observation lies at the heart of some of the more unethical aspects of contemporary urban (mal-) development, such as the increase in homelessness, inequality, decline of social housing, gentrification and so on. These are viewed as unavoidable consequences of global city competition;

(7) Competition between global cities generates other social costs. For example, as cities compete to host major events like the Olympics, the resulting impacts fall heavily and unevenly on residents, as analysed by Susan Fainstein in relation to London (Fainstein 2010).

The global city requires a constant supply of migrants and foreign workers absorbed into low-wage and informal labour markets in order to support higher layers of urban society (see Figure 6.2). Put simply, our current world system of cities functions on the basis of a steady supply of cheap labour. The world city system is predicated on the existence of urban hierarchies being played out and reinforced in rankings. The picture is one of unsustainable, inequitable, unaccountable and corporately dominated cities.

The competitive city orientation is both a predisposition to the brutal simplicity of winners and losers in the city markets and a predilection for winning against neighbouring cities, even locked into a global race for status and wealth. Both facets of the competitive city are increasingly being challenged. Patterns of urbanisation and economic power are ever shifting. While incumbents and wealthy residents are clearly advantaged in the competitive city, they are also affected in various ways by the negative impacts of urban growth that is uncontrolled, unregulated and unjust (more slums, poverty, growing inequality, pollution, health hazards etc.). Moreover, there is an increasing recognition even amongst competitive city advocates that everything depends upon avoiding catastrophic climate change, increasing resilience in the face of natural disasters and navigating a range of other existential challenges facing modern civilisation, not least the potential for devastating global pandemics such as Covid-19.

In the battle for rankings, city league tables increasingly position themselves as purportedly reflecting liveability when they are, in fact, also competitive in orientation and therefore do not give primacy to the fundamentals of equality, environmental sustainability or democratic representation. Better known examples include The Economist Intelligence Unit's (EIU) Global Liveability Index, the Mercer Quality of Living Ranking and the rather eclectic Monocle Quality of Life Survey (See Table 6.1). The first two focus primarily on the liveability needs of mobile employees of transnational corporations.

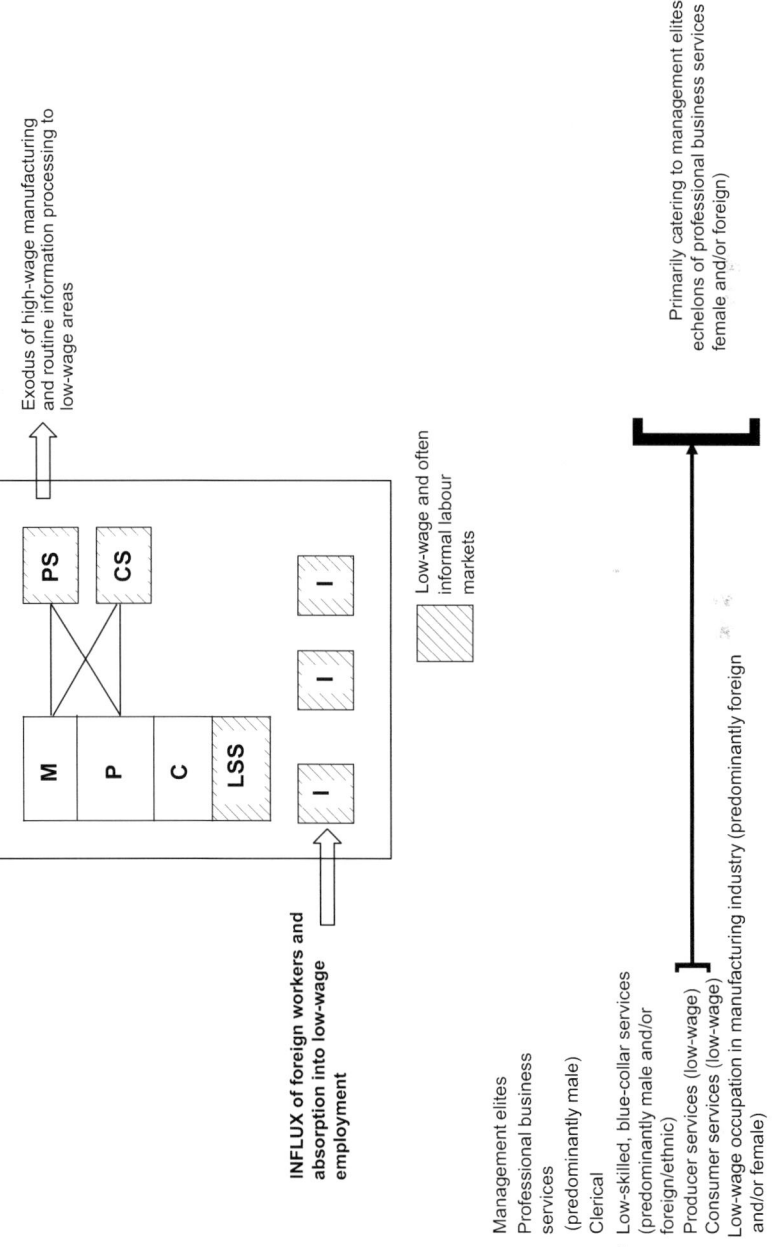

Exodus of high-wage manufacturing and routine information processing to low-wage areas

Primarily catering to management elites and upper echelons of professional business services (predominantly female and/or foreign)

INFLUX of foreign workers and absorption into low-wage employment

Low-wage and often informal labour markets

M   Management elites
P   Professional business
    services
    (predominantly male)
C   Clerical
LSS Low-skilled, blue-collar services
    (predominantly male and/or
    foreign/ethnic)
PS Producer services (low-wage)
CS Consumer services (low-wage)
I   Low-wage occupation in manufacturing industry (predominantly foreign
    and/or female)

FIGURE 6.2 Employment structure and functioning of a world city

Source: Friedmann (1986, p. 78).

TABLE 6.1 Top ten ranked cities in 2019

EIU Global Liveability Ranking	Mercer Quality of Living Ranking	Monocle Quality of Life Survey
Vienna	Vienna	Zurich
Melbourne	Zurich	Tokyo
Sydney	Vancouver	Munich
Osaka	Munich	Copenhagen
Calgary	Auckland	Vienna
Vancouver	Dusseldorf	Helsinki
Tokyo	Frankfurt	Hamburg
Toronto	Copenhagen	Madrid
Copenhagen	Geneva	Berlin
Adelaide	Basel	Lisbon

Even if we forgive the faulty core logic of such rankings and accept their findings, what next? The focus of international attention for investment to improve liveability is rarely placed on the least liveable cities such as Damascus, Dhaka, Port Moresby, Lagos, Tripoli, Karachi, Algiers and Harare. It is not difficult to imagine how it must feel for the mayor of each these cities when the EIU issues the annual press release declaring the most liveable city. Just take a second to think about the subsequent headlines in the Dhaka Tribune: 'Second least liveable city again, for six years in a row!' While Dhaka faces numerous challenges, including poor infrastructure, unregulated, rapid growth, congested roads, political instability and corruption (Institute of Governance 2012), the majority of Dhaka's residents gain nothing from existing rankings. That might be a very different story if rankings could be reoriented towards addressing systemic urban problems of inequality, environment and governance.

Orientation 2: urban liveability

An urban liveability orientation reflects the lived urban experience and substitutes the lens of competition with one of care and concern for quality of life. For example, research undertaken in Australia assessed urban liveability against a set of criteria that included walkability, access to public transport, open space, housing affordability, employment and the environments for food and alcohol (Arundel et al. 2017). The researchers found considerable spatial variability within cities. The attainment of policy standards differs significantly between city districts. In general, if you live in the outer suburbs of an Australian city then you are less likely to be well served in terms of amenities and infrastructure, compared to inner city residents. While these findings are not new, repeated empirical evidence is a means to raise concerns about disparity of quality of life due to variable infrastructure and services, as long as these inequalities continue to be rolled out.

Such work also contributes to a shift in language, rhetoric and institutional alignments around city priorities. It compliments platforms such as the Sustainable Development Goals (see Chapter 5). A key observation we draw from this work is the critical dependence on data availability resulting in limited applicability for localities where robust time series data on such indicators are unavailable. Another observation is that while indicators reveal useful trends and directions at the city level, they could if clumsily applied elsewhere inadvertently downplay lived experiences of residents across the city who face uneven circumstances across education, health, income, housing ownership and tenure, social status, legal status and so on (McArthur and Robin 2019). Any analysis of urban liveability, if it is to be truly representative, must therefore focus on inequality within and between cities and the underpinning causality. Globally, urgency for the adoption of such a conceptualisation has been popularly framed by the urban theorist, Mike Davis, who estimated that by 2050 there will be two billion people living in slums – equivalent to the entire population of Indonesia, Pakistan and the Philippines (Davis 2007).

Abrupt disparities in wealth and opportunity are unhealthy and make unliveable cities. While a competitive city orientation might shun action on poverty and sustain high crime and poor health outcomes as a result, in the liveable city orientation poverty has a distinct empirical and policy focus. For instance, the idea of the poverty line has substance and provides a basis for action. Data for 2018 reveals that 9.8 percent of the Indonesian population lived below the national poverty line (US$1.90 a day) (Asian Development Bank 2020). In Japan, it is set at US$30 per day (Nippon.com 2017) and, in 2016, one in six Japanese people (particularly the young and single mothers) were found to be living beneath this poverty line in conditions that are generally hidden from the rest of society (Osaki 2016). The poverty line in Australia for a single person is AUD$61 per day (US$40) and the proportion of Australians living below this line stood at 10.7 percent in 2017 (Henriques-Gomes 2019). Across the United States, see Figure 6.3, the poverty rate in 2018 was 11.8 percent, amounting to 38 million people (Semega et al. 2019). The point here is not to take issue with the complex and contestable methods of setting poverty lines, rather, it is that the idea and advancement of such a mechanism promotes debate, response and relevance to the idea of care and quality of life. While these figures are at national rather than city level, one can envisage how city poverty can promote attention to liveability, in ways that offer a stark contrast to the often shocking yet mundane encounters with extreme poverty in the competitive city – familiar and invisible to many passers-by (as we discussed in Chapter 1).

In the search for liveability, measurement tools, like all bows to evidence and empiricism, have the potential to hide as much as they reveal and emphasise. The decision to assess particular parameters of urban experience inevitably renders other parameters less important. The act of selection actually devalues and dims the visibility of everything else. The danger here is that when we measure

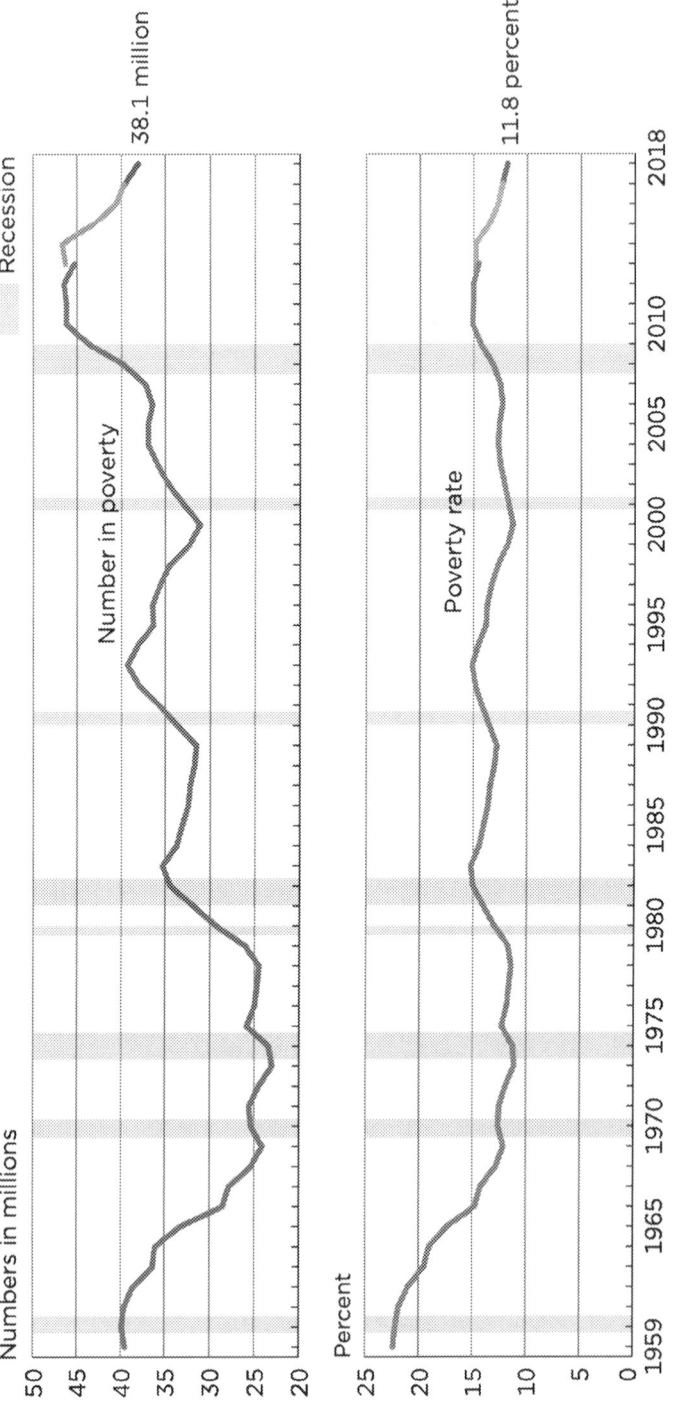

FIGURE 6.3   Number in poverty and poverty rate in the United States: 1959–2018

Source: Semega et al. (2019, p. 12).

Note: The data is derived from the Current Population Survey, *1960–*to 2019 *Annual Social and Economic Supplements*, U.S. Census Bureau.

liveability by readily available data we miss potentially important issues. At the same time, when we fail to make the invisible visible (as we discussed in Chapter 1) we may miss potential solutions that are currently unimaginable. As the American architect, Buckminster Fuller remarked in a 1983 interview:

> I think it's absolutely touch-and-go whether we're going to make it. But the point is, for me to tell you that you have an option to be optimistic . . . I am running into millions who don't know we have the option, because it's invisible, and I feel I have tremendous responsibility. So when people ask me to come and talk to them, I do my best to let them know they do have the option.
>
> *(Fuller 1983)*

A deceptively obvious (and rarely adopted) way to identify what it is about liveabilty that is important to urban residents is to simply ask them. In 2018–19, the London Sustainable Development Commission undertook research to investigate the concerns of Londoners aged 16–24, highlighting significance of poverty, housing affordability, homelessness and employment (Sloam and Ehsan 2019). An online survey undertaken as part of this research identified the top six priorities of young Londoners as knife crime, affordable housing, homelessness, air pollution, radicalisation, extremism and job protection. Key social issues were mental and physical health, crime, education and gender equality, while economic concerns were housing affordability, child poverty, jobs availability and cost of transport. According to the report 'for young Londoners, poverty and inequality were overarching issues that drove many concerns regarding their futures, impacting all the other areas addressed – from mental health, to knife crime, to community cohesion' (ibid., p. 15). Questions of inequality were also significant. Many respondents raised concerns about the growing wealth gap between the rich and poor in London. In addition, intergenerational equity was seen as problematic with one young female interviewee remarking:

> We're getting hit because this country has an ageing population, and it seems that the older generation are the ones that are making most of the policies, even though they are going to be the ones that are least affected for the least time.
>
> *(ibid., p. 62)*

The realities of post-Brexit London are far from the ideals of liveability, and in the darker aspects of isolationism and pandemic they resemble scenes from Alphonso Cuarón's *Children of Men*; a 2027 police state hunting down illegal immigrants and seeking to maintain some semblance of order in the midst of a collapsing world (Cuarón 2006). Aligned with competitive city agendas, fear is created and built into the functions of cities, in tandem with creeping privatisation of public spaces (Minton 2009), the creation of which require high security as 'pre-requisite of planning permission for all new developments' (Minton and Aked 2013). For your own protection, Londoners have an estimated 500,000 (and counting) CCTV cameras on the streets, with some four million across the UK (Caught on Camera 2020). Fear has enabled what Stephan Graham describes as *New Military Urbanism* (Graham 2013), where the once-humble police officer has taken on a paramilitary role and '[m]ilitary-style borders, fences and checkpoints around defended enclaves and 'security zones', superimposed on the wider and more open city, are proliferating.' Elsewhere '[t]hey are growing around strategic financial districts, embassy zones, tourist spaces, airport and port complexes, sport event spaces, gated communities and export processing zones' (Graham 2010, p. xxi). Jeremy Nemeth similarly observes the emergence of security zones around Wall Street in New York and other key districts in US cities (Nemeth 2010). Public concerns about police militarisation led the Obama administration to prohibit equipment made for the battlefield that is not appropriate for local police departments in January 2015 (Law Enforcement Equipment Working Group 2015). The prohibited equipment list included tracked armoured vehicles, weaponised aircraft, grenade launchers, firearms of .50-calibre or higher ammunition, bayonets and camouflage uniforms (Thomas 2016). These prohibitive measures were subsequently rescinded by the Trump Administration.

Public reaction to both urban militarisation and urban poverty share commonalities with climate change. Although these problems are hugely consequential in their implications for urban liveability, they are rendered unactionable and, ultimately, inevitable by existing social and political structures. The explanation that is insidiously promoted is that we can know something is true, but to act as if it is not (Marshall 2014, pp. 226–229). The excesses of wealth and partition in our cities are thus rendered invisible, materially, through the proliferation of private, gated sections of cities and of penthouse apartments in high-rise buildings across the city, as partitioned off alpha territories for the ultra-rich (Graham 2016; Glucksberg 2016; Minton 2017). They are also hidden financially – although Transparency International UK has revealed the darker side to the 40,000 properties in London registered to overseas companies in secrecy jurisdictions that ensure the owner's identity remains hidden (de Simone 2015), enabling London to become the world's number-one home to the fruits of corruption (Shenker 2016). They are also hidden by virtue of the portrayal of the competitive city as the normal, unchangeable, way of things; a state that is beyond question. In such circumstances, a concentration of super-rich in London owning a collective fortune estimated at £333 billion goes largely unquestioned (Hennig and Dorling 2015). The result is a great inversion with the core of London becoming gentrified (moving upmarket) and those who cannot afford rising house prices and rents forced outward (The Economist 2013). This raises profound questions about liveability in global cities like London.

Orientation 3: fragile cities

Whether mesmerised by the competitive power of the top 100 world cities, or concerned with quality of life and the measurability of liveability, there is, it seems, little room for considering the plight of the thousands of cities that are much less prepared to cope with crises (Muggah 2014a). Virtually all cities are fragile, but the intensity of their fragility varies considerably (Muggah 2014b). Fragility occurs when city authorities are unwilling or unable to deliver basic services and this triggers a rupture in the city's social contract (de Boer 2015; Muggah 2016). On a continuum between spatial and social integration on one end and economic, cultural and ethnic polarisation on the other (Calame and Charlesworth 2009), fragility increases in line with growing polarisation.

The term fragile in this context sits alongside a range of pejorative adjectives applied to cities including feral, failed, divided, fragmented and violent. The 2018 ranking of the 50 most violent cities prepared by the Citizens' Council for Public Security based in Mexico puts Tijuana and Acapulco (Mexico) in the top two places and Caracas (Venezuela) in third with 138, 110 and 100 murders per 100,000 population, respectively. A total of 15 cities from Mexico and 14 from Brazil were included in this ranking, together with six from Venezuela and four from the United States: St. Louis, Baltimore, Detroit and New Orleans (Citizens' Council for Public Security 2020).

Fragility is invariably associated with both an inability to cope or be resilient and a current reality of poverty, inequality and violence. While these are inter-related manifestations of life in many cities of the Global South (Salahub et al. 2018), it is no accident that the prevalence of economic crime is closely associated with disadvantaged suburbs and inner cities, whether they are in the Global North or South. That inequality is prevalent and largely unaddressed is a matter of justice. No one votes to be poor; hence the fact that poverty is endemic in cities around the globe suggests a justice gap and a fundamental flaw in representational politics. Ever present and yet often largely invisible, corruption of power and diversion of collective wealth is institutionalised, sophisticated and built into social and political fabric via legitimised power structures within and across cities.

Fragility manifests as insecurity and urban mega-slums are both hostile places to live and potential havens for hostile non-state actors (United States Army 2014). As such, they are of interest to both local elites and to established rich nations globally. Their security apparatus is accordingly attuned to monitoring fragile states and cities. Robert Muggah and his team at the Igarape Institute have collated data on 2,137 cities worldwide with populations over 250,000 (representing 55 percent of the world's population). The indicators used cover population growth, unemployment, inequality, pollution, climate risks, homicide and exposure to terrorism but exclude exposure to natural disasters (Igarape Institute 2020). The Fragile States Index provides a starting point for assessing fragile cities (Fund for Peace 2017). This is a national-scale risk assessment tool, employing a diverse range of social, economic, political and military indicators

on demographic pressures, refugees, uneven economic development, group griev-
ances, human flight and brain drain, poverty, state legitimacy, public services,
human rights and the rule of law, security apparatus, factionalised elites and
external interventions (from foreign military occupation to foreign assistance
and the presence of peacekeepers).

In this context, Richard Norton, a US Naval War College scholar, coined
the term feral city (Norton 2003). As shown in Table 6.2, a city is considered to
have gone feral when government control ceases to exist or is restricted to certain
zones, when the economy functions like a black market, where services are intermittent or only available to those who can afford to pay, and where security is
assured through private means or by paying for protection. According to Norton
the picture of the city is that of:

> [a] mosaic, and like an artist's mosaic it can be expected to contain more
> than one color. Some healthy cities function with remarkable degrees of

TABLE 6.2 How cities go feral

	Government	Economy	Services	Security
Healthy (Green)	Enacts effective legislation, directs resources, controls events in all portions of the city all the time. Not corrupt.	Robust. Significant foreign investment. Provides goods and services. Possesses stable and adequate tax base.	Complete range of services, including educational and cultural, available to all city residents.	Well-regulated by professional, ethical police forces. Quick response to wide spectrum of requirements.
Marginal (Yellow)	Exercises only a patchwork or diurnal control. Highly corrupt.	Limited/ no foreign investment. Subsidised or decaying industries and growing deficits.	Can manage minimal level of public health, hospital access, potable water, trash disposal.	Little regard for legality/ human rights. Police often matched/ stymied by criminal peers.
Going Feral (Red)	At best has negotiated zones of control; at worst does not exist.	Either local subsistence industries or industry based on illegal commerce.	Intermittent to non-existent power and water. Those who can afford to will privately contract.	Non-existent. Security is attained through private means or paying protection.

Source: Norton (2003, p. 101).

corruption. Others, robust and vital in many ways, suffer from appalling levels of criminal activity. Even a city with multiple 'red' categories is not necessarily feral – yet. It is the overall pattern and whether that pattern is improving or deteriorating over time that give the overall diagnosis.

(Norton 2003, p. 103)

Subsequently, Norton elucidated a diagnostic tool that assists the 'evaluation of whether a city is trending towards becoming feral' as presented in Table 6.3 (Norton 2010, p. 57).

Muggah's contention is that there are many more fragile cities outside of the so-called fragile states as within them. He concludes that turbo-urbanisation is a major determinant of fragility, with cities exhibiting population growth at a rate of above 4 percent most at risk. A particular indicator identified is a high proportion of younger people in the overall population – a youth bulge. About 85 percent of the world's youth live in fragile states and in particular circumstances high-youth populations connect to high levels of youth unemployment and street crime (Muggah 2015).

Addressing fragility is thus a focus for widely contrasting reasons – to name three: for the lives of those on the receiving end; for the risks to those in richer places and for the ethical rightness at stake. The posterchild of the turnaround city is Colombia's second largest city – Medellín. Dominated by drug cartels in the 1980s and 1990s, the past decade has seen Medellín embark on an extraordinary transformation from a murder hotspot to an apparent civic beacon, making headlines around the world (Moss 2015). Medellín was one of the most world's dangerous cities until the cartel was broken by coordinated government, police and public action. Transparency and accountability were central to the tenure of Mayor Aníbal Gaviria (2010–2015), enabling investment in new public infrastructure and food security programmes (Andrews 2014). Respectful engagement at the grassroots with a vulnerable and shocked post-trauma populace proved essential in the transformation of the city into an increasingly safe and successful urban centre. The Medellín project revolved around six key measures:

(1) A social urbanism strategy focussed on projects empowering and resourcing poorer communities;
(2) Upgrading public transport infrastructure and enhancing connectivity, including construction of cable car lines to connect poor, remote communities;
(3) Enforcement of a Green Belt around the city, providing opportunities for local food production;
(4) Establishing a network of enterprise development zones;
(5) Participatory budgeting to give residents more say in how the city spends its funds and
(6) Supporting arts-based cultural initiatives to reclaim public spaces.

(McLaren and Agyeman 2015, pp. 191–196)

TABLE 6.3 Diagnosing ferality

Governance	Economy	Services	Security	Civil Society
Green Level (No Danger)				
Enacts effective legislation. Appropriately directs resources. Controls events in all portions of the city day and night. Corruption detected and punished.	Robust. Significant foreign investment. Provides goods and services. Possesses a stable and adequate tax base.	Complete range of services, including educational and cultural, available to all city residents.	Well regulated by professional, ethical police forces. Quick response to wide spectrum of requirements.	Rich and robust. Constructive relationships with government.
Yellow Level (Marginal)				
Exercises only patchwork or diurnal control. Highly corrupt.	Limited or no foreign investment. Subsidised or decaying industries. Growing deficits. Most foreign investment is quickly removable.	Can manage minimal level of public health, hospital access, potable water, trash disposal.	Little regard for legality/human rights. Police often matched/stymied by criminal 'peers'.	Relationships with government are confrontational. Civil society organisations fill governmental voids.
Red Level (Becoming Feral)				
At best has only negotiated zones of control.	Local subsistence industries or industries based on illegal commerce. Some legitimate business interests may be presented based on profit potential.	Intermittent to non-existent power and water. Those who can afford to will privately contract or allow NGO to provide.	Non-existent. Security is attained through private means or paying protection.	Civil society fractured along clan/ethnic/other lines. Local elites in control. Security-oriented civil society organisations may be criminal.

Source: Norton (2010, p. 57).

Perhaps the single most important observation to make on the Medellín case, however, is that it contradicts Muggah's account, in particular, regarding the causal link drawn between street crime and the presence of unemployed youth, especially young men. 'The youth' is in fact hugely diverse and socially constituted, rather than a homogenous and discrete group and is not in fact responsible for gang violence. To the extent that there is a link, this group is merely a symptom. The *causes* are numerous and situated and include poverty, inequality, aggressive hierarchies, victimisation, loss of identity and ability for self-actualisation and ontological security. In acknowledging this, at least to some degree, Medellín was able to develop programmes that addressed these causes, rather than simply problematising youth. The truth is that cities and their demographics are always mosaics and as such defy simple categorisation. Research is needed to prise apart the orientations presented in this chapter and to seek out models and practices that offer hope in promoting solutions to poverty, inequality, failing sustainability and lack of accountability in urban governance systems. Our collective challenge, therefore, is to ensure that decisions, whether in cities of the industrialised or emergent worlds, are made with ethical city considerations at the forefront.

Orientation 4: fightback cities – fearless, rebellious and/or ethical

Just as fragile cities are a direct result of the dominance of hierarchies of power spawning powerlessness and poverty, rather than of a wayward or deviant youth, so, fearless cities are people and places where the incumbency of power is laid bare and where everything is on the table in the pursuit of a just society. David Harvey's term Rebel Cities evokes an overt activist orientation for cities at the centre of anti-capitalist resistance that are focussed upon developing processes for enhancing local democracy and economic autonomy (Harvey 2013). He explains how capitalism itself has gone feral with corrupt politicians and bankers, CEOs and hedge fund managers looting global wealth and promoting tax avoidance. Juxtaposing this with rioting and looting on the streets of London in 2011, he regards such reactions as mimicking what corporate capital is doing to the planet (ibid., p. 56). The urban uprising is promoted as a means to portray publicly the failure of liberal democracies to address issues of exclusion and related economic inequalities (Dikec 2017). The overt anti-capitalist stance of rebel and fearless cities is taken up further in Chapter 8.

By making visible the invisible, the ethical city illuminates the failings that competitive cities engender and that are manifest in fragile cities. It also provides a principled framework for action, accepting that behaviours and actions are bounded by situation, structure and agency (or lack thereof). Fearlessness is hard to come by, but even in small doses provides many more options and reasons to be optimistic in the face of seemingly high odds and fixed disadvantage. Fundamentally, many cities are defined by the struggle to cope with the capital accumulation and labour impacts of globalisation, and while all cities face their own particular circumstances, common measures can ameliorate the challenges of the fragile city and engender liveability and fearlessness in the ethical city. Here are some initial propositions adapted from Muggah (2015):

(1) Engage with vulnerable communities within and across cities, providing solidarity and support as an antidote to competitive and divisive logics;
(2) Focus on the causes of crime and disaffection rather than upon symptomatic stereotypes such as criminal youth and create programmes that include, engage and build identity and social status;
(3) Invest in social cohesion and mobility, for example, through enhancements to public transportation systems, support for inclusive public spaces and pro-poor policies;
(4) Reform multilateral systems of solidarity to operate both through and beyond the state, engaging between cities for mutual aid and networking;
(5) Engage with liveability and other related indicators as a means to promote awareness of the root causes of city inequalities and to build democratic, accountable engagement that empowers a critical and active orientation;
(6) Promote critical, fearless engagement through urban politics and a pro-activist orientation.

References

Andrews, J. (2014) Interview with Aníbal Gaviria Correa, Mayor of Medellin, Colombia, *Cities Today*. [Online]. Available: http://cities-today.com/interview-with-anibal-gaviria-correa-mayor-of-medellin-colombia/ [Last accessed: 10 May 2020].

Arundel, J., Lowe, M., Hooper, P., Roberts, R., Rozek, J., Higgs, C. and Giles-Corti, B. (2017) *Creating Liveable Cities in Australia*. Melbourne: Centre for Urban Research, RMIT University. [Online]. Available: https://cloudstor.aarnet.edu.au/plus/index.php/s/CJ4t5N3SFCOZTWP [Last accessed: 20 May 2020].

Asian Development Bank (2020) *Poverty Data: Indonesia*. [Online]. Available: www.adb.org/countries/indonesia/poverty [Last accessed: 2 March 2020].

Badhwar, N.K. (2009) The milgram experiments, learned helplessness, and character traits, *Journal of Ethics 13*: 257–289. https://doi.org/10.1007/s10892-009-9052-4.

Calame, J. and Charlesworth, E. (2009) *Divided Cities: Belfast, Beirut, Jerusalem, Mostar and Nicosia*. Philadelphia: University of Pennsylvania Press.

Calvino, I. (2004) On invisible cities, *Columbia: A Journal of Literature and Art 40*: 177–182. [Online]. Available: www.jstor.org/stable/41808770 [Last accessed: 7 May 2020].

Castells, M. (2000) *The Rise of the Network Society*, 2nd Edition. Oxford: Blackwell Publishing.

Caught on Camera (2020) *How Many CCTV Cameras in London?* [Online]. Available: www.caughtoncamera.net/news/how-many-cctv-cameras-in-london/ [Last accessed: 2 March 2020].

Citizens' Council for Public Security (2020) *Ranking 2018*, in Spanish. [Online]. Available: www.seguridadjusticiaypaz.org.mx/seguridad/1564-boletin-ranking [Last accessed: 2 March 2020].

Cuarón, A. (2006) *Children of Men*, Science Fiction Drama. New York: Universal Pictures.

Davis, M. (2007) *Planet of Slums*. Brooklyn: Verso.

de Boer, J. (2015) Resilience and the fragile city, *Stability: International Journal of Security and Development 4* (1). http://doi.org/10.5334/sta.fk.

de Simone, M. (2015) Corruption on your doorstep: How corrupt capital is used to buy property in the UK. [Online]. Available: www.transparency.org.uk/publications/corruption-on-your-doorstep/ [Last accessed: 28 April 2020].

Dikec, M. (2017) *Urban Rage: The Revolt of the Excluded*. New Haven and London: Yale University Press.

Dobbs, R., Smit, S., Remes, J., Manyika, J., Roxburgh, C. and Restrepo, A. (2011) *Urban World: Mapping the Economic Power of Cities*, McKinsey Global Institute. [Online]. Available: www.mckinsey.com/~/media/McKinsey/Featured%20Insights/Urbanization/Urban%20world/MGI_urban_world_mapping_economic_power_of_cities_full_report.ashx [Last accessed: 10 May 2020].

The Economist (2013) *Mapping Gentrification: The Great Inversion: A Closer Look at the Gentrification of London*. [Online]. Available: www.economist.com/blighty/2013/09/09/the-great-inversion [Last accessed: 1 April 2020].

Fainstein, S.S. (2010) *The Just City*. Ithaca: Cornell University Press.

Friedmann, J. (1986) The world city hypothesis, *Development and Change 17*: 69–83. https://doi.org/10.1111/j.1467-7660.1986.tb00231.x.

Fuller, R.B. (1983) *Interview: Only Integrity Is Going to Count*. [Online]. Available: www.bfi.org/about-fuller/resources/articles-transcripts/only-integrity-going-count [Last accessed: 28 April 2020].

Fund for Peace (2017) Fragile and conflict affected states. [Online]. Available: http://fundforpeace.org/global/what-we-do/fragile-and-conflict-affected-states/[Last accessed: 1 April 2020].

Giddens, A. (1984) *The Constitution of Society: Outline of the Theory of Structuration*. Berkeley: University of California Press.

Glucksberg, L. (2016) A view from the top: Unpacking capital flows and foreign investments in prime London, *City 20* (2): 238–255. https://doi.org/10.1080/13604813.2016.1143686.

Graham, S. (2010) *Cities under Siege: The New Military Urbanism*. London: Verso Books.

Graham, S. (2013) Foucault's boomerang: The new military urbanism, *Open Democracy*. [Online]. Available: www.opendemocracy.net/en/opensecurity/foucaults-boomerang-new-military-urbanism/ [Last accessed: 3 March 2020].

Graham, S. (2016) *Vertical: The City from Satellites to Bunkers*. London: Verso Books.

Hall, P. (1966) *The World Cities*. New York: McGraw-Hill.

Harvey, D. (2013) *Rebel Cities: From the Right to the City to the Urban Revolution*. London: Verso Books.

Hennig, B. and Dorling, D. (2015) London, the British Isles and the super-rich, *London-mapper: A Social Atlas of London*. [Online]. Available: http://london.worldmapper.org/super-rich/ [Last accessed: 1 April 2020].

Henriques-Gomes, L. (2019) Poverty is rising again in Australia and expert cites welfare changes as likely cause, *The Guardian*. [Online]. Available: www.theguardian.com/australia-news/2019/jul/29/poverty-is-rising-again-in-australia-and-expert-cites-welfare-changes-as-likely-cause [Last accessed: 2 March 2020].

Huxley, A. (1932) *Brave New World*. London: Vintage Books, published in 2007.

Igarape Institute (2020) *Fragile Cities: How Is City Fragility Spread around the World?* [Online]. Available: https://public.tableau.com/profile/igarape#!/vizhome/FragileCities/FragileCities [Last accessed: 1 April 2020].

Institute of Governance (2012) *State of Cities: Urban Governance in Dhaka*, BRAC University, Dhaka, Bangladesh. [Online]. Available: http://dspace.bracu.ac.bd/xmlui/bitstream/

handle/10361/2055/SOC%20Report-%2005-07-12.pdf?sequence=1&isAllowed=y [Last accessed: 10 May 2020].

International Organization for Migration (2015) *World Migration Report 2015- Migrants and Cities: New Partnerships to Manage Mobility*, Geneva, IMO. [Online]. Available: https://publications.iom.int/system/files/wmr2015_en.pdf [Last accessed: 1 April 2020].

Law Enforcement Equipment Working Group (2015) *Recommendations Pursuant to Executive Order 13688, Federal Support for Local Law Enforcement Equipment Acquisition.* [Online]. Available: https://bja.ojp.gov/sites/g/files/xyckuh186/files/publications/LEEWG_Report_Final.pdf [Last accessed: 3 March 2020].

Marshall, G. (2014) *Don't Even Think about It: Why Our Brains Are Wired to Ignore Climate Change.* New York: Bloomsbury.

McArthur, J. and Robin, E. (2019) Victims of their own (definition of) success: Urban discourse and expert knowledge in the liveable city, *Urban Studies, Debates Paper*: 1–18. https://doi.org/10.1177/0042098018804759.

McLaren, D. and Agyeman, J. (2015) *Sharing Cities: A Case for Truly Smart and Sustainable Cities.* Cambridge, MA: MIT Press.

Minton, A. (2009) *Ground Control: Fear and Happiness in the Twenty-First-Century City.* London: Penguin.

Minton, A. (2017) *Big Capital: Who Is London For?* London: Penguin.

Minton, A. and Aked, J. (2013) *Fortress Britain: High Security, Insecurity and the Challenge of Preventing Harm*, Prevention Working Paper, New Economics Foundation. [Online]. Available: www.annaminton.com/single-post/2016/03/21/New-report-Fortress-Britain [Last accessed: 29 April 2020].

Moss, C. (2015) Medellín, Columbia: A miracle of reinvention, *The Guardian*. [Online]. Available: www.theguardian.com/travel/2015/sep/19/medellin-colombia-city-not-dangerous-but-lively [Last accessed: 1 April 2020].

Muggah, R. (2014a) *How to Protect Fast Growing Cities from Failing.* [Online]. Available: www.ted.com/talks/robert_muggah_how_to_protect_fast_growing_cities_from_failing [Last accessed: 10 May 2020].

Muggah, R. (2014b) Deconstructing the fragile city: Exploring insecurity, violence and resilience, *Environment & Urbanization 26* (2): 345–358. https://doi.org/10.1177/0956247814533627.

Muggah, R. (2015) Fixing fragile cities: Solutions for urban violence and poverty, *Foreign Affairs*. [Online]. Available: www.foreignaffairs.com/articles/africa/2015-01-15/fixing-fragile-cities [Last accessed: 28 April 2020].

Muggah, R. (2016) How fragile are our cities? *World Economic Forum*. [Online]. Available: www.weforum.org/agenda/2016/02/how-fragile-are-our-cities [Last accessed: 27 April 2016].

Nemeth, J. (2010) Security in public space: An empirical assessment of three US cities, *Environment and Planning A 2*: 2487–2507. https://doi.org/10.1068/a4353.

Nippon.com (2017) Japan's poverty rate remains well above OECD average. [Online]. Available: www.nippon.com/en/behind/l10354/japan%27s-poverty-rate-remains-well-above-oecd-average-news.html [Last accessed: 2 March 2020].

Norton, R.J. (2003) Feral cities, *Naval War College Review 56* (4): 97–106. [Online]. Available: https://digital-commons.usnwc.edu/nwc-review/vol56/iss4/8/ [Last accessed: 10 May 2020].

Norton, R.J. (2010) Feral cities: Problems today, battlefields tomorrow, *Marine Corps University Journal 1* (1): 51–77. [Online]. Available: www.usmcu.edu/Portals/218/HD%20MCUP/MCUP%20Pubs/DA%20JOURNAL%20FINAL.pdf?ver=2018-11-21-074835-670 [Last accessed: 5 June 2020].

Osaki, T. (2016) Hidden poverty growing under Abe, particularly among young and single mothers, *Japan Times*. [Online]. Available: www.japantimes.co.jp/news/2016/04/26/national/social-issues/hidden-poverty-growing-abe-particularly-among-young-single-mothers/ [Last accessed: 28 April 2020].

Salahub, J.E., Gottsbacher, M. and de Boer, J. (eds.) (2018) *Social Theories of Urban Violence in the Global South: Towards Safe and Inclusive Cities, Routledge Studies in Cities and Development*. London and New York: Taylor and Francis Group.

Sassen, S. (1991) *The Global City: New York, London, Tokyo*. Princeton and Oxford: Princeton University Press.

Semega, J., Kollar, M., Creamer, J. and Mohanty, A. (2019) *Income and Poverty in the United States: Current Population Reports*, US Department of Commerce. [Online]. Available: www.census.gov/library/publications/2019/demo/p60-266.html [Last accessed: 2 March 2020].

Shenker, J. (2016) Which are the most corrupt cities in the world? *The Guardian*. [Online]. Available: www.theguardian.com/cities/2016/jun/21/which-most-corrupt-cities-in-world [Last accessed: 4 October 2020].

Sloam, J. and Ehsan, R. (2019) *Young Londoners' Priorities for a Sustainable City*, London Sustainable Development Commission, Greater London Authority, London. [Online]. Available: www.london.gov.uk/sites/default/files/young_londoners_report_final_0.pdf [Last accessed: 29 April 2020].

Thomas, A. (2016) *What You Need to Know about Executive Order 13688*. [Online]. Available: www.policeone.com/jag/articles/what-you-need-to-know-about-executive-order-13688-7PjAN7e7PZBrCzAz/ [Last accessed: 3 March 2020].

United States Army (2014) *Megacities and the United States Army: Preparing for a Complex and Uncertain Future*. Arlington: United States Army. [Online]. Available: www.army.mil/e2/c/downloads/351235.pdf [Last accessed: 29 April 2020].

Upton, C.L. (2009) Virtue ethics and moral psychology: The situationism debate, *Journal of Ethics 13*: 103–115. https://doi.org/10.1007/s10892-009-9054-2.

7

RELENTLESS DISRUPTION

The inferno of the living is not something that will be; if there is one, it is what is already here, the inferno where we live every day, that we form by being together. There are two ways to escape suffering it. The first is easy for many: accept the inferno and become such a part of it that you can no longer see it. The second is risky and demands constant vigilance and apprehension: seek and learn to recognize who and what, in the midst of the inferno, are not inferno, then make them endure, give them space.

Italo Calvino, Invisible Cities, p. 148

Utopian and dystopian cities

Los Angeles, as portrayed in Ridley Scott's science fiction film *Blade Runner*, was set in a dystopian future (Scott 1982). The director envisioned a city in 2019 that was polluted, climate-changed, crowded and absent of nature. There was a constant police presence and those who could afford it were migrating to off-world colonies (Dunn et al. 2014; Barrett 2015). It was a city dominated by huge corporations capable of replicating human and animal life. The Los Angeles of 2019 was not as Scott predicted. Many elements portrayed in the film, however, are manifest in different cities across the world. In our collective imaginary we share a rich history of urban dystopic storytelling that has a sobering sense of plausibility – Lang's classic *Metropolis*, Huxley's *Brave New World* or Wells' *Shape of Things to Come* (Lang 1927; Huxley 1932; Wells 1933). Whether it is inequality extremes with the wealthy living in high-rise buildings, while humble workers dwell underground maintaining machines or totalitarian scenarios where people are psychologically manipulated from birth and escape reality through drugs (soma), supposed fiction is perilously close to today's urban lived experience for some on our planet. Within the sci-fi genre, Zamyatin's novel *We* envisages a city constructed entirely of glass to facilitate mass surveillance, with citizens referred to by numbers (D-503, 1–330, O-90) rather than names (Zamyatin 1921). It is a world without privacy, chillingly prescient of the heavily surveilled CCTV cities of today and the privacy invaded, digitally connected online presence where huge oligopolistic companies like Facebook and Google (and anyone accessing their data about us) can know everything including our patterns of social media behaviour. This surveillance capitalism (Zuboff 2019) is characterised by the commodification of personal information.

Dystopian visions are all the more disturbing when they appear to be increasingly real and normalised. At the other end of the urban futures spectrum, utopian visions have also provided their share of disruption, whether we are talking about Le Corbusier's Ville Radieuse, Ebenezer Howard's Garden City or Frank Lloyd Wright's Broadacre. They are precursors to contemporary urban megaprojects, including smart cities that may turn out to be not so smart, like Songdo, South Korea or ecological cities that turn out to be not so sustainable, like Masdar in the United Arab Emirates (Newitz and Stamm 2014; Cugurullo and Ponzini 2019). Utopian visions and blueprints are subject to the unpredictability of urban change and the wicked character of societal problems (Rittel and Webber 1973; Vandenbroeck 2012; Hui 2015). There is a wild nature to our cities shaped by patterns of migration, slums, unemployment, homelessness, crime, air traffic congestion and air pollution (Castells 1976; Steele 2019). When faced with such immense challenges we do need utopian thinking, but it must be guided by realism in order to navigate and overcome existing structures of power and politics (Bregman 2017). While there is no lack of imagination embodied in our future-oriented visions and dreams, there remains an expansive gulf between vision and reality. This may be a blessing in disguise since historically some of the worst excesses associated with elitist visions have had to be curbed by the activism of grassroot movements of urbanists and urbanites motivated by notions of justice and ethics (Castells 1976; Harvey 2013).

The many urban deficits require that we redouble our efforts to identify, monitor and tackle risks, ensuring that society becomes increasingly aware of intractable uncertainties and adaptive in the face of relentless change (Elahi 2011). We have to cope with constant disruption, including the implications of emerging technologies like artificial intelligence, automation and hyper-monitored cities. There are many negative consequences concomitant with late neoliberal global capitalism producing increasingly short-term, episodic work and manic consumption. The hacking of democracy and enduring impacts of post-GFC austerity are co-products of these circumstances (as we discussed in Chapters 3 and 4). Despite the apparently increasing ubiquity of global systems of production, exchange and consumption, the negative consequences are experienced highly unevenly – not just between rich and poor, but also across climate zones, across geopolitical power blocs and across demographics. Some cities struggle to deal with rapid population growth while others face ageing and/or population decline and economic decay. With citizens largely regarded as producers and consumers, livelihoods come and go as economic disruptions dictate. Cities become sites variously of success and failure, attracting and repelling labour accordingly, while mobility is highly privileged and politicised. Across these uneven urban terrains, a new terminology has emerged encompassed in the disruptive genre – smart, autonomous, artificial, intelligent and digital. Attempts to create a universal language for contemporary cities belies their unevenness and the consequences for urban residents. If there is a generality it is that cities are different, yet they all share tendencies towards contradictions; beautiful boulevards and ugly slums; high technology and urban decay; highly functional spaces and in-between wastelands; thousands of close-living social beings many of whom are highly isolated and lonely (Sennett 2018, pp. 293–294). The so-called liveable cities, whether Vienna, Melbourne or Osaka, are great places to live if you are well off, but tough if you are in casual labour market trying to make ends meet and living in the marginal rental housing market. Even the most advanced cities in the world face uncertainty about their future and struggle to cope with prolonged disruptions.

Mega-trends

Mega-trends, including technological determinism, urbanisation, globalisation and demographic change, shape socio-economic structures across the globe (Gebalska 2017). The shift towards increased inequality reduced accountability and fractured democracy are key urban mega-trends with profound consequences. To these we can add the climate and ecological crisis, global interconnectedness and pandemic diseases (Commission for Future Generations 2013; Cribb 2017; Laybourn-Langton et al. 2019). In some countries, Japan is a good example, we have witnessed slowing fertility rates, population growth, GDP growth per capita and technological innovation with potentially positive implications for the planet in terms of reducing pressures (Pearce 2011; Dorling 2020). In others, there has been frenetic casualisation of work, fuelled by digital innovation, hyper-consumption and a general accelerationism where things are likely to get worse faster and where there is limited vision of the future beyond disruption (Shaviro 2015). Where the impacts of unfettered post-industrial restructuring are coupled with this casualised consumption, the impacts are profound; for example, rapid manufacturing decline in the UK and a shift to banking and financial services (Kitson and Michie 2014), together with deregulation and demise of the public sector as a source of stable employment. As employment growth slows and quality of employment declines, the situation is compounded by falling rates of net in-migration of economically active age groups and rising rates of poverty (Pike et al. 2016). No path forward is presented beyond relentless disruption.

The UK is by no means unique in this regard. Many cities across the United States, for example, have been losing manufacturing jobs since the 1970s. Indeed, over seven million US manufacturing jobs, about 38 percent of the national manufacturing base, were lost between 1980 and 2009, with 61 percent of these in metropolitan areas (Friedhoff et al. 2010, p. 2). While there has been growth in jobs in advanced services, education, health, transportation, warehousing and tourism, the decline in manufacturing employment has been accompanied by falling employment rates and quality, including fewer hours worked and lower wages. The social impact is compounded by a rise in drug use and economic crime rates in areas hardest hit by manufacturing decline, with implications for these places in terms of their ability to attract future employment (Charles et al. 2018, p. 3). The next phase could involve these cities losing their remaining manufacturing, retail and service industries to automation. At the same time, service, transportation and warehouse jobs are shifting to artificial intelligence. The digital society has transformed our social and political interactions but not necessarily in a good way. While rapid technological change has been all-pervasive and ongoing, it has not impacted on work hours, long commutes, gig economy jobs, insecure work contracts, stagnating pay, credit card debt, high rents and mortgage repayments, university loans and disappearing pensions (Srnicek and Williams 2015; Graeber 2018).

Even prior to Covid-19, economic changes in the United States in recent decades had impacted greatly on many Americans with data showing 50 million lived in economically distressed areas (Economic Innovation Group 2018, p. 4). A distressed economy is defined in terms of severely restricted access to education, housing and employment as well as high levels of poverty. Another 56 million people were considered to be at risk of slipping into distress, living from pay cheque to pay cheque. Another important fact is that the United States was already ranked as a second-tier country according to the 2018 Social Progress Index (see www.socialprogress.org) suggesting that it was failing to meet citizens' needs compared to other advanced industrial economies (ranked twenty-fifth, behind Portugal, Singapore, Slovenia and Italy). In the UK (ranked thirteenth), this segment of the population is referred to as the precariat – a new class of disempowered individuals who face job insecurity and low wages without pensions, paid holidays and other benefits and unable to rely on social support systems such as unemployment and sickness benefits (Standing 2011). This is reminiscent of conditions people faced in the late twentieth century, before reforms that brought the 40-hour week, paid holidays and national insurance. Progress made during half a century of relative prosperity following World War II seems to have rapidly given way as a neoliberal trapdoor opened up beneath the working and middle classes.

This dismal picture has not been limited to post-industrial Western society. From Brazil to the Philippines, there has been an unfolding arena of elite interests driving rapid, rampant and unsustainable urbanisation. The imbalanced economic benefits of urban projects with attendant uneven individual and communal rights have brutally confronted the collective right of decision-making. Lack of provision for accessible physical and social infrastructure was too often excused by the pace of change, as cities could not keep up with the pace of economic transformation. Rapid innovation creates redundant technologies, vacant sites of production and discarded labour and skills employed, leaving large segments of cities and their populations economically isolated, alienated, unemployed and facing increased inequalities (Verdis and Burdett 2013).

Returning to the dystopian/utopian binary, barring some new intervention, two possible endgames appear plausible. On one hand, according to the UK philosopher Nick Land, *accelerationism* from the perspective of the political right involves continued intensification of capitalism leading to technological singularity – a point in time where technological growth becomes uncontrollable, irreversible and unfathomable – revealing an extreme form of the neoliberal city (Noys 2017, pp. 2–3). On the other hand, *accelerationism* for the political left involves redirection of the tools of capitalist modernity and new technologies for liberatory and emancipatory ends. Under this narrative, the aim of progressives is to regain control of modernity, to be future-oriented and to think big (Srnicek and Williams 2015, p. 183). Whatever happens it is clear that the city will be the terrain over which much of this struggle takes place. The question is whether those who live and work in the constantly disrupted city will be able to gain some degree of control and therefore ameliorate the impact of the powerful array of forces aligned against them or whether they will remain as victims (Noys 2017, p. 14). The urban terrain for this struggle is set and next we explore two key theatres of battle: (1) automation and artificial intelligence (AI) and (2) zero-carbon urban futures.

Automation and artificial intelligence

Automation and artificial intelligence have been described as part of the Fourth Industrial Revolution (Schwab 2015; Pitroda and Miailhe 2017). If we are alert to the mode of and motivation for the adoption of these technologies then it is a moot point as to whether they represent a fantastic opportunity to ignite productivity or deliver an employment apocalypse. Automation has been with us since the start of the First Industrial Revolution and continues to advance at a rapid pace. Between 2010 and 2015, sales of industrial robots in Asia grew by 888,000, and there were 2.6 million industrial robots in operation worldwide, mainly in the automotive, electronics, metals, chemicals and plastics sectors. Recent data shows the largest proportion of robot sales is to China (43 percent), South Korea (24 percent) and Japan (22 percent) (International Federation of Robotics 2016). The impacts of automation have been accelerating with one study indicating 47 percent of jobs in the United States were at risk from automation (Frey and Osborne 2013, p. 38). More conservative estimates from the OECD suggest only 14 percent of jobs at risk, while another 32 percent could change significantly (Nedelkoska and Quintini 2018, p. 115). This is already evident in many areas. For instance, there are still immigration officers at most airports, but their role is to check on passports rejected by the automated systems or to deal with any individuals identified as problematic by the technology. Likewise, at supermarkets we increasingly witness only one or two staff members around to help customers who encounter problems with self-service tills. A visit to an Amazon distribution centre reveals a working environment almost completely dominated by automation.

In 2019, a study by the Brookings Institution revealed that 25 percent of US employment (36 million jobs) faces high exposure to automation, while 36 percent faces medium exposure (Muro et al. 2019, p. 31). These divergent figures can be highly confusing for policymakers, urban planners and society in general, making it difficult to craft appropriate responses (Osborne and Frey 2018). Another study examined a total of 702 occupations and found that workers who were exposed to automation tended to have lower levels of education and (typically) lower incomes (Frey and Osborne 2013). The researchers predicted that those most at risk included workers in transportation and logistics, as well as the bulk of office and administrative support workers and labour in production occupations. In the United States alone that amounted to 3.5 million truck drivers (Day and Hait 2019).

Moreover, research by the Center for an Urban Future revealed that 12 percent of the workforce in New York State could be replaced with technology that already exists (Chaban 2018). This includes 155,000 food preparers and servers, 124,000 accounting and bookkeeping clerks and 181,000 delivery truck and bus drivers. The temptation to automate these jobs will gain increased attention after the events of 2020. It has already been estimated in Australia that 25 percent to 46 percent of work activities will be automated by 2030 across nine sectors − (1) retail, (2) administration, (3) construction, (4) manufacturing, (5) accommodation and food services, (6) transport and warehousing, (7) healthcare, (8) professional services and (9) education. This would translate to between 3.5 million and 6.5 million full-time equivalent positions displaced over the next decade (Taylor et al. 2019, p. 6). The Brookings Institution study, mentioned earlier, indicated that routine predictable physical and cognitive tasks will be the most vulnerable. It also revealed that automation risks vary across regions and cities, with the US Heartland (19 central states), smaller rural communities and smaller metropolitan areas worst affected (Muro et al. 2019; Frank et al. 2018). Male workers appear to be more vulnerable to future automation than women. Young workers between the ages of 16 and 24 face a high average exposure to automation of around 49 percent. Figure 7.1 illustrates the average automation potential by state in the United States and Figure 7.2 provides the projections on levels of automation for local government areas in Australia for 2030. The political implications for already polarised nations recovering from Covid-19 are all too clear.

Pre-Covid-19 it was estimated that AI would deliver US$13 trillion globally in additional economic activity by 2030 – 16 percent more than would otherwise be anticipated (Taylor et al. 2019, p. 13). China has been consistently investing in this sector with a total output value for the Chinese AI industry estimated at $147 billion in 2030. In the period 2010–2014, the US led the AI patent applications (15,317) with China in second place (8,410), followed by Japan and South Korea (UNESCAP 2018, p. 4). The pervasiveness of AI is undeniable, and

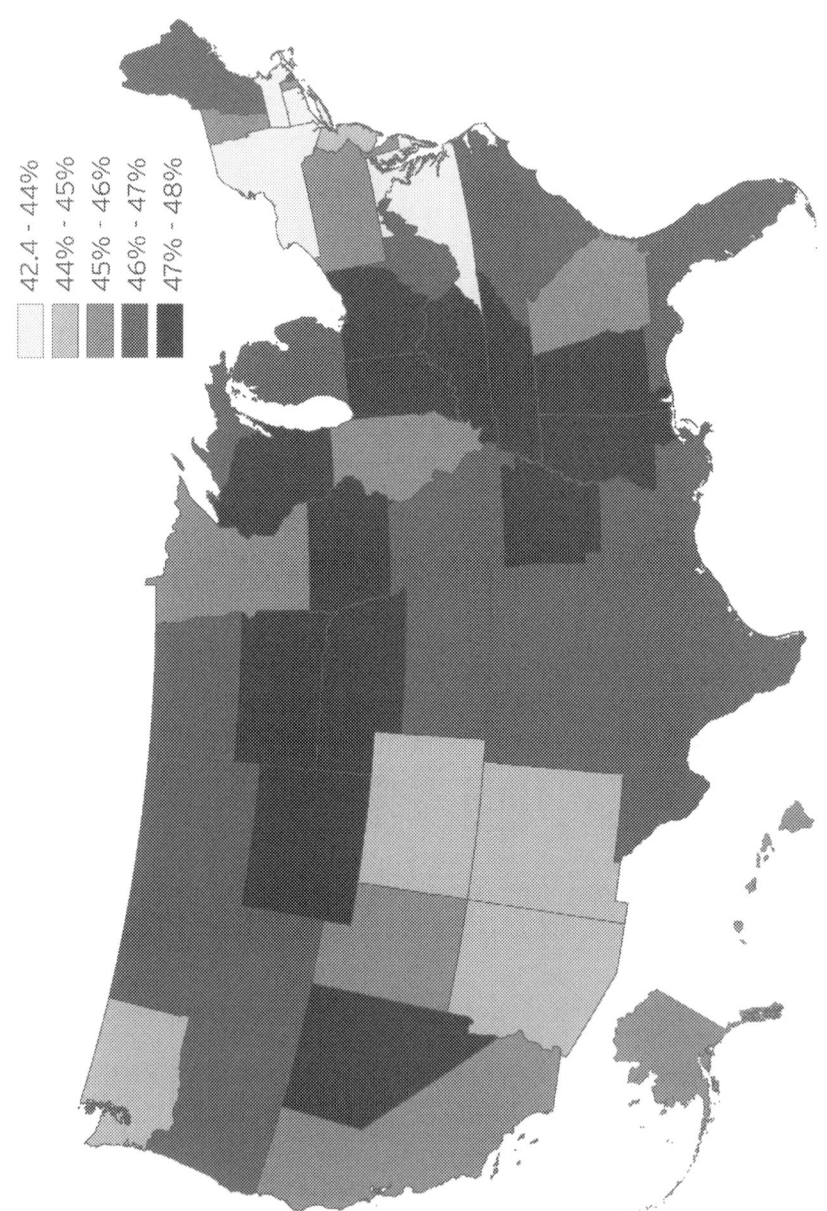

FIGURE 7.1 Average automation potential by US State in 2016

Source: Muro et al. (2019, p. 38).

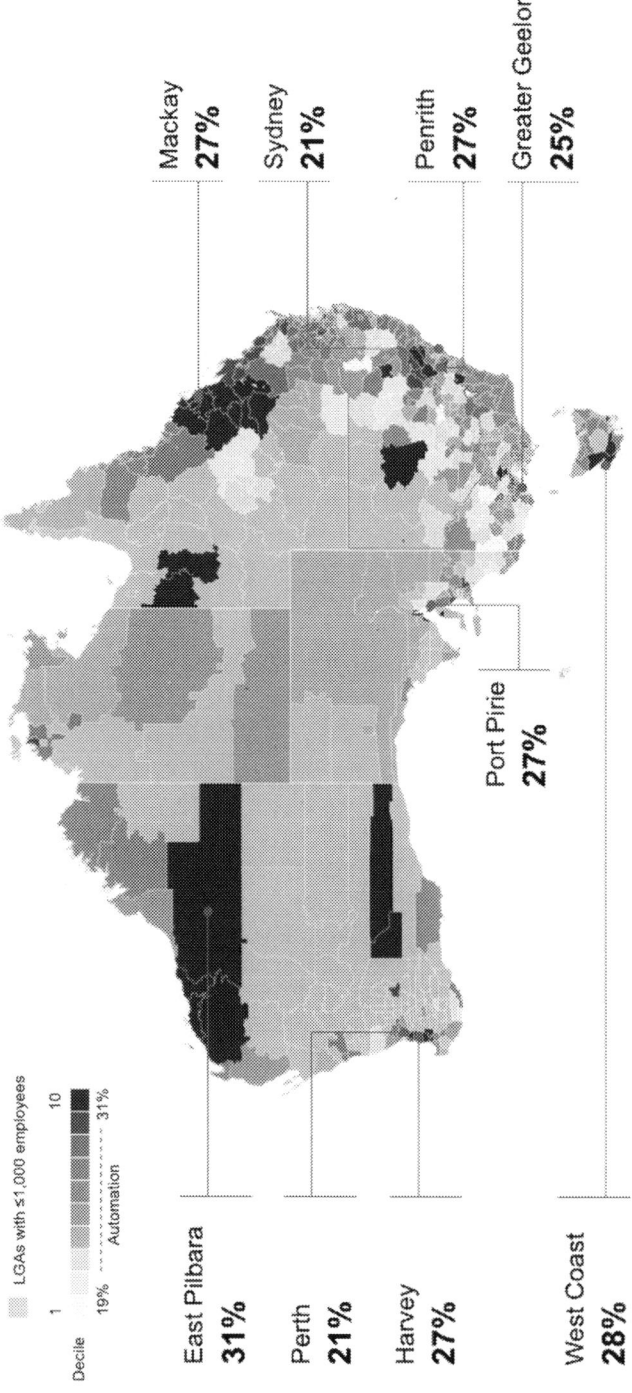

FIGURE 7.2   Average automation potential by Australian local government areas in 2030

any urban scale response faces familiar dilemmas. A dominant world-view positions these as forces for good, albeit with unfortunate side effects. While nothing could be further from the truth, the dominance of global capital flows and productivity (GDP) measures as unquestioned directives for policy mean it is a brave government that actually stands up and says 'no' to these technologies. Saying no automation or no AI is not regarded as an irrational choice (Taylor et al. 2019), despite the inevitable pain to workers affected in the absence of a safety net. Moreover, the predicted capital and productivity benefits are repeatedly recounted as truth-telling. In the case of Australia, for example, the increase in productivity from automation is estimated to add AUD$1.1 trillion to AUD$4 trillion to the economy by 2030. To realise these benefits (although it is never spelled out exactly who will net these windfalls and how unevenly they will fall) we are encouraged to take/accept five major strategies: (1) embrace economic growth and technological change; (2) promote a constant learning mindset; (3) facilitate smoothening adjustments; (4) reduce worker hardships and (5) mitigate harsh local impacts through future-proofing of vulnerable regional economies (Muro et al. 2019, p. 68). Resisting automation is futile because if you are not riding this technological wave then your economy will drown. Regulation is similarly regarded as of modest value in countering negative consequences (Cranshaw 2013). Indeed, trying to plan beyond a nod to the aforementioned strategies is regarded as counterintuitive. The only game in town is to bet everything on winning the AI and automation sweepstakes and deal with the consequences later (rather than plan for them in advance). In this context, South Korea tops the list of favourite automation-ready countries, followed by Germany, Singapore and Japan, with the United States and Australia at the ninth and tenth positions, respectively. (Economist Intelligence Unit 2018, p. 10).

Zero-carbon urban futures

Amidst the climate emergency, the Intergovernmental Panel on Climate Change (IPCC) has pivoted towards cities in its calls for action, reflecting the fact that they account for the majority of emissions and that significant mitigation action can be undertaken at the municipal level. The IPCC's pleas have also become increasingly urgent because 'in model pathways with no or limited overshoot of 1.5°C, global net anthropogenic CO_2 emissions [must] decline by about 45 percent from 2010 levels by 2030' (IPCC 2018, p. 12). The situation remains complicated regarding intergenerational and inter-regional inequities, however, because of the meshed nature of emissions from cities of the Global North and Global South (C40 2018) and the enormous legacy of destruction and consumption being left by the baby boomers across the Global North. While New York, Paris, London and other wealthy cities have been implementing significant measures to reduce their production-related carbon emissions (the stuff we make), research has revealed that their consumption-related emissions (the stuff we buy from overseas) actually triple their carbon footprints. Deciding what we count and how to attribute it is fundamentally an ethical issue. The change required is obviously opposed to business as usual, contemporary structures and values systems (Marshall 2014).

The battle to decarbonise is well underway, but many obstacles need to be overcome. While climate denial remains well-funded and orchestrated, it will inevitably fail and, thankfully, many fossil fuels will be left in the ground as stranded assets. The IPCC 2018 Special Report requires removal of CO_2 emissions by technical measures in order to achieve negative emissions (see Figure 7.3). There are many questionable assumptions around the technical, economic and social viability of such measures (Anderson and Peters 2016). Moreover, viability of carbon sinks is unknown. This implies risks beyond those already known to be occurring and multiplying – extreme weather events, reduced water availability, lower crop yields, sea-level rise and coral reef system destruction (Schleussner et al. 2016). As the global centres of economic activity and carbon demand, cities are shaping up as the sites in which low-carbon economies are emerging, central to the attainment of deep decarbonisation scenarios (Floater and Rode 2014; Bazaz et al. 2018; ICLEI 2018). They will also increasingly be seen as havens for climate refugees. In this context, the mode of urban growth from now on is going to be critical in achieving carbon neutrality.

The dominant discourse around low-carbon cities shifted over the past decade from 'if and can' to 'how and when'. This does not mean the low-carbon city is a foregone conclusion; far from it. We often hear calls for new models of urban growth based on compactness, connected infrastructure and coordinated governance. Stockholm is put forward as an example of this model since, in the period between 1993 and 2010, the city was able to reduce emissions by 35 percent while increasing economic output by 41 percent (Floater and Rode 2014, p. 34). The economic and social benefits of achieving rapid decarbonisation of UK cities have also been examined, with 16 pathways related to energy efficiency in buildings, transportation and waste management identified (Gouldson et al. 2018, p. 7). These pathways are typically peppered with major concerns around the limitations and observations that our current fossil fuel-based urban infrastructure is not in line with the need to avoid the 1.5°C overshoot (Smith et al. 2019). They also include attendant calls for strong and enforced national policy frameworks. Predictions tend to be top-down, mechanical and based on technologically formulated options designed to provide confidence to investors and policymakers rather than rooted in social realities of urban change (for instance, should we be driving our cars everywhere? What is essential and non-essential travel?). Accordingly, the costs always appear miraculously feasible, amounting to around 1–2 percent of GDP in the case of the UK (Committee on Climate Change 2019, p. 8). Once again, the way in which these costs are distributed is typically not raised.

Chiming in with a new discourse of low-carbon cities are the many city networks pledging collaboration, self-help and forging ahead in the absence of national and global leadership. One among many is the C40 cities initiative

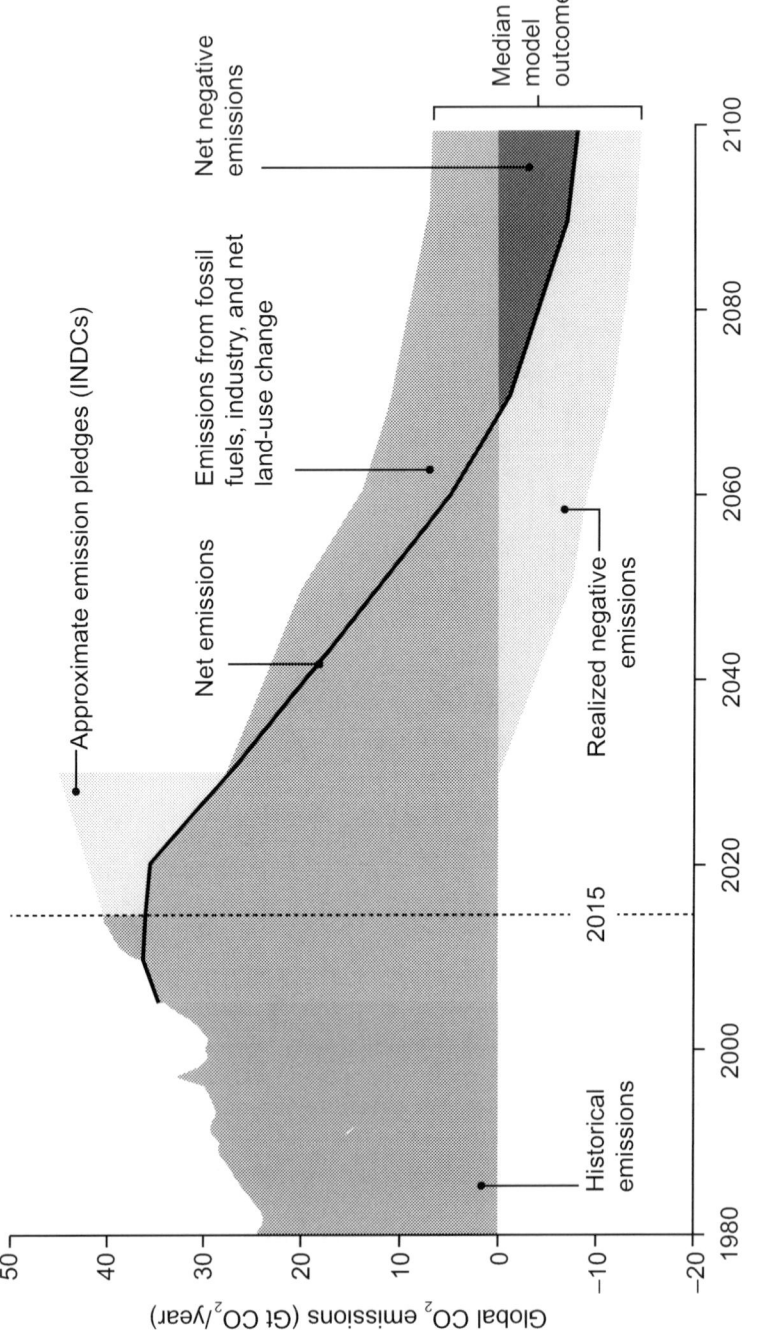

FIGURE 7.3 Deep decarbonisation and negative emissions scenarios compared to emission pledges from the Paris Agreement

Source: Anderson and Peters (2016, p. 182).

which has the 'Deadline 2020' banner under which a concerted assault on emissions was orchestrated over a four-year period between 2016 and 2020. The goal was to reduce average C40 per capita emissions from 5.1 to 4.86 tCO2e/capita between 2015 and 2020, despite a predicted 10 percent increase in total population (Hurst and Clement-Jones 2017, p. 46). Unfortunately, the report's authors subsequently admitted to errors in their methodology that led to overestimation of business as usual emissions trajectories in the report. This raises a very important concern about our ability to accurately measure city greenhouse gas emissions (Barrett and DeWit 2017). Early city-scale emissions estimates were flawed and only rectified in 2014 with the Global Protocol for Community-Scale GHG Emissions Inventories prepared by the World Resources Institute, ICLEI and the C40 Cities Climate Leadership Group (Fong et al. 2014).

Measurement of emissions remains challenging for many cities. A 2018 investigation of 79 cities revealed that inclusion of consumption emissions resulted in a 60 percent increase compared to those calculated using the GPC methodology (C40 Cities 2018). Moreover, when Copenhagen's emissions were measured on a consumption basis, they turned out to be four-to-five times higher than on a territorial basis (Dahal and Niemela 2017). GHG emission accounting at the city level is still very much work-in-progress, but thankfully the methodologies are improving rapidly (Broekhoff et al. 2019). It is indicative of the complexity embodied in the transition to a zero-carbon urban future as we struggle to understand what it means, who is involved, how it should unfold and how to measure or recognise progress (Luque-Ayala 2018). But that does not mean that zero-carbon cities are unattainable.

Disruption and business-unusual solutions

Whether it is the spectre of automation and AI or the steep slope towards climate neutrality, there exists an impressive array of current conventional solutions. In 2019, the OECD issued a set of recommendations requiring that AI-related developments should respect environmental sustainability, human rights and democracy. Governments are encouraged to take steps to ensure a fair transition for workers impacted by AI deployments, providing access to training and new employment opportunities (OECD 2019). In the UK, the Royal Society for the Encouragement of Arts, Manufactures and Commerce (RSA) has been exploring various scenarios for the labour market in 2035. Four futures of work scenarios were identified as follows: (1) *Big Tech Economy* – a world where technologies develop at a rapid pace; (2) *Precision Economy* – future of hyper surveillance; (3) *Exodus Economy* – characterised by economic slowdown and (4) *Empathy Economy* – a future of responsible stewardship where technologies and their implications are carefully managed (Dellot et al. 2019, p. 7). Several recommendations to smoothen the transition towards an Empathy Economy were put forward, including rebalancing the burden of the tax system to increase the taxes on unearned income (e.g. dividends and inheritances) while reducing that on earned income; launching a future of work alliance and introducing a Charter on Ethical Technology Investments. Key questions to address with each scenario and policy interventions are presented in Table 7.1. The report authors argue that:

> if we continue to ignore the possibility of alternative futures, and talk only of mass automation versus business as usual, we risk putting the liveli-hoods of workers in jeopardy. Rising inequality, growing suppression in the workplace, stagnant wages, heightened discrimination and bias, and deepening geographic division could all come to pass if we do not become more responsible custodians of technology. Experience tells us we cannot be complacent.
>
> *(ibid.: p. 54)*

The ethical dimensions of technology are highlighted in Dellot *et al.'s* scenarios presented in Table 7.1. The proposal is for investors to steer technology in more ethical directions and for new auditing tools to be applied systematically across the economy in order to more effectively promote benevolent technologies (ibid., p. 58). This is particularly relevant to smart city technologies with methodolo-gies already evolving to map and address their ethical dimensions (Greenfield 2013; Kitchin 2016). Key ethical concerns related to smart technologies include: erosion of privacy; lack of public consent during project implementation; limited oversight; data ownership issues; unanticipated repurposing of data; biases in data and AI systems; digital divides and differential access to digitally mediated services (Kitchin 2019a, 2019b). One response involves the notion of technologi-cal sovereignty as implemented in Barcelona – where people, communities and their local governments retain overall control of technological platforms (Tie-man 2017).

In line with our ethical city principles, action on climate change must be inte-grated with action on inequality and accountability; it must be just and equitable. There is even less room for complacency here. For example, in 2015 the Interna-tional Labour Organisation issued guidelines for a just transition towards envi-ronmentally sustainable economies (International Labour Organization 2015). Calls for a just transition (making social interventions needed to secure workers and their communities' jobs and livelihoods as we shift away from fossil fuels) have been issued in Poland, France and through the Green New Deal in the United States (Moreno et al. 2019). These guidelines are in reaction to decades of laissez-faire policies; dumping mining workers on the dole when coal prices or other pressing elite imperatives dictate. The just transition will need to occur in a world characterised by extreme patterns of wealth inequality; not only are the richest responsible for the majority of global emissions, but they also hold the strongest cards at the labour relations table. As a result, these transitions can be expected to be non-linear, at times accelerating and at other times slowing down and even stalling (Geels 2005). It seems inevitable that more radical and

TABLE 7.1 Pressing questions and priority interventions for each scenario

	Questions to Address	*Priority Interventions*
Big Tech Economy	How can the power of tech giants be contained? How can we prevent unemployment from soaring? Can and should certain technologies be outlawed? How should we manage future mergers and acquisitions?	Introduce a comprehensive technology sentry system. Give every worker a technology inheritance through a UK sovereign wealth fund. Update competition law to reflect the needs of workers as well as consumers.
Precision Economy	How can the contingent workforce (including the self-employed) be protected? How do we ensure worker monitoring is proportionate? How can we improve the collection, storage and use of worker data? How can worker rating systems be fair and transparent?	Pilot personal learning accounts (available to all workers). Establish a new welfare settlement for the self-employed. Clarify employment status law and strengthen the enforcement of worker rights. Introduce a new right to data portability (critical for platform work).
Exodus Economy	How can the pace of technology development and adoption be accelerated? How can the migration of workers in search of jobs be facilitated? How can we promote alternative economic institutions and new union models? How can the unemployed and underemployed be supported?	Address the flaws in Universal Credit and scale trials of a Universal Basic Income. Amend legislation to make it easier to join a union (e.g. digital ballots). Promote alt union models built on new power principles.
Empathy Economy	How do we promote self-regulation among tech giants and firms using their technology? How can growth of empathy sectors (i.e. care work, teaching, personal sport professionals) be facilitated? How do we prevent jobs in the empathy sectors from being commoditised? How can the emotional demands of labour be contained?	Modernise recruitment practices in the tech sector. Introduce a Charter for Ethical Technology Investments. Establish a prize challenge for technology vetting tools. Establish a union dedicated to tech workers.

Source: Dellot et al. (2019, p. 56).

new alliances will be needed, perhaps borne out of the misery of dispossessed workers in the AI transition and/or of dispossessed islander communities fleeing climate change. We turn to the somewhat more optimistic topics of responses to ethical city building over the next two chapters.

References

Anderson, K. and Peters, G. (2016) The trouble with negative emissions: Reliance on negative-emission concepts locks in humankind's carbon addiction, *Science 354* (6309): 182–183. https://doi.org/10.1126/science.aah4567.

Barrett, B.F.D. (2015) Ethical cities are the future, *Our World*. [Online]. Available: http://ourworld.unu.edu/en/ethical-cities-are-the-future [Last accessed: 6 February 2020].

Barrett, B.F.D. and DeWit, A. (2017) This is why we cannot rely on cities alone to tackle climate change, *The Conversation*. [Online]. Available: https://theconversation.com/this-is-why-we-cannot-rely-on-cities-alone-to-tackle-climate-change-82375 [Last accessed: 27 April 2020].

Bazaz, A., Bertoldi, P., Buckeridge, M., Cartwright, A., de Coninck, H., Engelbrecht, F., Jacob, D., Hourcade, J., Klaus, I., de Kleijne, K., Lwasa, S., Markgraf, C., Newman, P., Revi, A., Rogelj, J., Schultz, S., Shindell, D., Singh, C., Solecki, W., Steg, L. and Waisman, H. (2018) *Summary for Urban Policymakers: What the IPCC Special Report on 1.5C Means for Cities*, Indian Institute for Human Settlements. [Online]. Available: www.ipcc.ch/site/assets/uploads/sites/2/2018/12/SPM-for-cities.pdf [Last accessed: 10 May 2020].

Bregman, R. (2017) *Utopia for Realists and How We Can Get There*. London and New York: Bloomsbury.

Broekhoff, D., Erickson, P. and Piggot, G. (2019) *Estimating Consumption-Based Greenhouse Gas Emissions at the City Scale: A Guide for Local Governments*. Stockholm: Stockholm Environment Institute. [Online]. Available: www.sei.org/wp-content/uploads/2019/03/estimating-consumption-based-greenhouse-gas-emissions.pdf [Last accessed: 10 May 2020].

C40 Cities (2018) *Consumption-Based GHG Emissions of C40 Cities*, C40 Cities Climate Leadership Group. [Online]. Available: www.c40.org/researches/consumption-based-emissions [Last accessed: 27 April 2020].

Calvino, I. (1972) *Invisible Cities*. London: Vintage Books, published in 1997.

Castells, M. (1976) *The Wild City*. Kapitalistate: Working Papers on the Capitalist State: 2–30. [Online]. Available: www.marxists.org/history/usa/pubs/kapitalistate/Kapitalistate4&5.pdf [Last accessed: 10 May 2020].

Chaban, M.A.V. (2018) *State of Work: The Coming Impact of Automation on New York*, Center for an Urban Future. [Online]. Available: https://nycfuture.org/research/state-of-work-automation-impact-on-new-york-state [Last accessed: 8 April 2020].

Charles, K.K., Hurst, E. and Schwartz, M. (2018) *The Transformation of Manufacturing and the Decline of U.S. Employment*, National Bureau of Economic Research (NBER) Working Paper No. 24468. [Online]. Available: www.nber.org/papers/w24468 [Last accessed: 27 April 2020].

Commission for Future Generations (2013) *Now for the Long Term: The Report of the Oxford Martin Commission for Generations*. Oxford: University of Oxford. [Online]. Available: www.oxfordmartin.ox.ac.uk/downloads/commission/Oxford_Martin_Now_for_the_Long_Term.pdf [Last accessed: 10 May 2020].

Committee on Climate Change (2019) *Net Zero: The UK's Contribution to Stopping Global Warming, UK Government*. [Online]. Available: www.theccc.org.uk/wp-content/uploads/2019/05/Net-Zero-The-UKs-contribution-to-stopping-global-warming.pdf [Last accessed: 27 April 2020].

Cranshaw, J. (2013) *Whose "City of Tomorrow" Is It? On Urban Computing, Utopianism, and Ethics*, UrbComp'13: The 2nd ACM SIGKDD International Workshop on Urban Computing, August 11–13, 2013, Chicago, IL. [Online]. Available: SSRN: https://ssrn.com/abstract=2289964 [Last accessed: 27 April 2020].

Cribb, J. (2017) *Surviving the 21st Century: Humanity's Ten Great Challenges and How We Can Overcome Them*. Cham: Springer International Publishing.

Cugurullo, F. and Ponzini, D. (2019) The transnational smart city as urban eco-modernisation, the case of Masdar City in Abu Dhabi, in Karvonen, A., Cugurullo, F. and Caprotti, F. (eds.) *Inside Smart Cities: Place, Politics and Urban Innovation*. London and New York: Routledge.

Dahal, K. and Niemela, J. (2017) Cities' greenhouse gas accounting methods: A study of Helsinki, Stockholm, Copenhagen, *Climate 4* (31). https://doi.org/10.3390/cli5020031.

Day, J.C. and Hait, A.W. (2019) *Number of Truckers at All-Time High*, United States Census Bureau. [Online]. Available: www.census.gov/library/stories/2019/06/america-keeps-on-trucking.html [Last accessed: 27 April 2020].

Dellot, B., Mason, R. and Wallace-Stephens, F. (2019) *The Four Futures of Work: Coping with Uncertainty in the Age of Radical Technologies*. London: RSA. [Online]. Available: www.thersa.org/discover/publications-and-articles/reports/the-four-futures-of-work-coping-with-uncertainty-in-an-age-of-radical-technologies [Last accessed: 10 May 2020].

Dorling, D. (2020) *Slowdown: The End of the Great Acceleration: And Why It's Good for the Planet, the Economy, Our Lives*. New Haven and London: Yale University Press.

Dunn, N., Cureton, P. and Pollastri, S. (2014) *Future of Cities: A Visual History of the Future*. London: Government Office for Science. [Online]. Available: https://assets.publishing.service.gov.uk/government/uploads/system/uploads/attachment_data/file/360814/14-814-future-cities-visual-history.pdf [Last accessed: 10 May 2020].

Economic Innovation Group (2018) *From Great Recession to Great Reshuffling: Charting a Decade of Change Across American Communities: Findings from the 2018 Distressed Communities Index*. [Online]. Available: https://eig.org/wp-content/uploads/2018/10/2018-DCI.pdf [Last accessed: 27 April 2020].

Economist Intelligence Unit (2018) *The Automation Readiness Index: Who Is Ready for the Coming Wave of Automation?* London, New York, Hong Kong, Geneva and Dubai: The Economist. [Online]. Available: www.automationreadiness.eiu.com/static/download/PDF.pdf [Last accessed: 10 May 2020].

Elahi, S. (2011) Here be dragons . . . exploring the unknown unknowns, *Futures 43* (2): 196–201. https://doi.org/10.1016/j.futures.2010.10.008.

Floater, G. and Rode, P. (2014) *Cities and the New Climate Economy: The Transformative Role of Global Urban Growth*, NCE Cities Paper 01. London: LSE Cities. [Online]. Available: https://lsecities.net/wp-content/uploads/2014/12/The-Transformative-Role-of-Global-Urban-Growth-01.pdf [Last accessed: 27 April 2020].

Fong, W.K., Sotos, M., Doust, M., Schultz, S., Marques, A. and Deng-Beck, C. (2014) *Greenhouse Gas Protocol: Global Protocol for Community-Scale Greenhouse Gas Emission Inventories: An Accounting and Reporting Standard for Cities*. [Online]. Available: https://ghgprotocol.org/sites/default/files/standards/GHGP_GPC_0.pdf [Last accessed: 8 April 2020].

Frank, M.R., Sun, L., Cebrian, M., Youn, H. and Rahwan, I. (2018) Small cities face greater impact from automation, *Journal of the Royal Society Interface 15* (139). http://dx.doi.org/10.1098/rsif.2017.0946.

Friedhoff, A., Wail, H. and Wolman, H. (2010) *The Consequences of Metropolitan Manufacturing Decline: Testing Conventional Wisdom.* Washington: Metropolitan Policy Program, Brookings Institution. [Online]. Available: www.brookings.edu/research/the-consequences-of-metropolitan-manufacturing-decline-testing-conventional-wisdom/ [Last accessed: 10 May 2020].

Frey, C.B. and Osborne, M.A. (2013) *The Future of Employment: How Susceptible Are Jobs to Computerisation,* Oxford Management Programme on Technology and Employment, Oxford: Oxford University. [Online]. Available: www.oxfordmartin.ox.ac.uk/downloads/academic/The_Future_of_Employment.pdf [Last accessed: 8 April 2020].

Gebalska, B.A. (2017) Using global trends as catalysts for city transition, *Procedia Engineering 198*: 600–611. https://doi.org/10.1016/j.proeng.2017.07.114.

Geels F.W. (2005) The dynamics of transitions in socio-technical systems – A multilevel analysis of the transition pathway from horse-drawn carriages to automobiles, (1860–1930), *Technical Analysis and Strategic Management 17* (4): 445–476. https://doi.org/10.1080/09537320500357319.

Gouldson, A., Sudmant, A., Khreis, H., and Papargyropoulou, E. (2018) *The Economic and Social Benefits of Low-Carbon Cities: A Systematic Review of the Evidence.* London and Washington: Coalition for Urban Transitions. [Online]. Available: https://newclimateeconomy.report/workingpapers/wp-content/uploads/sites/5/2018/06/CUT2018_CCCEP_final.pdf [Last accessed: 10 May 2020].

Graeber, D. (2018) *Bullshit Jobs: A Theory.* New York: Allen Lane.

Greenfield, A. (2013) *Against the Smart City.* [Online]. Available: https://urbanomnibus.net/2013/10/against-the-smart-city/ [Last accessed: 10 May 2020].

Harvey, D. (2013) *Rebel Cities: From the Right to the City to the Urban Revolution.* London: Verso Books.

Hui, J.C. (2015) Planning ethics in the age of wicked problems, in I. Management Association (ed.) *Business Law and Ethics: Concepts, Methodologies, Tools, and Applications.* Hershey: IGI Global. doi:10.4018/978-1-4666-8195-8.ch003.

Hurst, T. and Clement-Jones, A. (2017) *Deadline 2020: How Will Cities Get the Job Done,* Report by C40 and Arup. [Online]. Available: www.c40.org/researches/deadline-2020 [Last accessed: 8 April 2020].

Huxley, A. (1932) *Brave New World.* London: Vintage Books, published in 2007.

ICLEI (2018) *Multilevel Climate Action: The Path to 1.5 Degrees.* Bonn: ICLEI. [Online]. Available: http://e-lib.iclei.org/wp-content/uploads/2018/12/cCR-report-web.pdf [Last accessed: 10 May 2020].

International Federation of Robotics (2016) *World Robotics Report.* [Online]. Available: https://ifr.org/ifr-press-releases/news/world-robotics-report-2016[Last accessed: 27 April 2020].

International Labour Organization (2015) *Guidelines for a Just Transition towards Environmentally Sustainable Economies and Societies for all.* Geneva: ILO. [Online]. Available: www.ilo.org/wcmsp5/groups/public/-ed_emp/-emp_ent/documents/publication/wcms_432859.pdf [Last accessed: 10 May 2020].

IPCC (2018) *Global Warming of 1.5°C. An IPCC Special Report on the Impacts of Global Warming of 1.5°C above Pre-Industrial Levels and Related Global Greenhouse Gas Emission Pathways, in the Context of Strengthening the Global Response to the Threat of Climate Change, Sustainable Development, and Efforts to Eradicate Poverty.* Edited by V. Masson-Delmotte,

P. Zhai, H.O. Pörtner, D. Roberts, J. Skea, P.R. Shukla, A. Pirani, W. Moufouma-Okia, C. Péan, R. Pidcock, S. Connors, J.B.R. Matthews, Y. Chen, X. Zhou, M.I. Gomis, E. Lonnoy, T. Maycock, M. Tignor, T. Waterfield. [Online]. Available: www.ipcc.ch/site/assets/uploads/sites/2/2019/06/SR15_Full_Report_High_Res.pdf [Last accessed: 10 May 2020].

Kitchin, R. (2016) The ethics of smart cities and urban science, *Philosophical Transactions A, 377*: 2083. https://doi.org/10.1098/rsta.2016.0115.

Kitchin, R. (2019a) The ethics of smart cities: Using big data and AI to manage cities creates many ethical issues, but efforts to address these concerns can be just as contentious, *Raidió Teilifís Éireann*. [Online]. Available: www.rte.ie/brainstorm/2019/0425/1045602-the-ethics-of-smart-cities/ [Last accessed: 10 May 2020].

Kitchin, R. (2019b) The ethics of smart cities, *Raidió Teilifís Éireann*. [Online]. Available: www.rte.ie/brainstorm/2019/0425/1045602-the-ethics-of-smart-cities/ [Last accessed: 10 May 2020].

Kitson, M. and Michie, J. (2014) *The Deindustrial Revolution: The Rise and Fall of UK Manufacturing 1870–2010*, Centre for Business Research, Working Paper No. 459. Cambridge: University of Cambridge. [Online]. Available: www.cbr.cam.ac.uk/fileadmin/user_upload/centre-for-business-research/downloads/working-papers/wp459.pdf [Last accessed: 10 May 2020].

Lang, F. (1927) *Metropolis*, Science Fiction Drama. Babelsberg, Potsdam: UFA.

Laybourn-Langton, L., Rankin, L. and Baxter, D. (2019) *This Is a Crisis: Facing Up to the Age of Environmental Breakdown: Initial Report*. Edinburgh: Institute for Public Policy Research. [Online]. Available: www.ippr.org/files/2019-11/this-is-a-crisis-feb19.pdf [Last accessed: 10 May 2020].

Luque-Ayala, A., Marvin, S. and Bulkeley, H. (2018) *Rethinking Urban Transitions: Politics in the Low Carbon City*. Abingdon and New York: Routledge.

Marshall, G. (2014) *Don't Even Think about It: Why Our Brains Are Wired to Ignore Climate Change*. New York and London: Bloomsbury.

Moreno, E., Krause, D. and Stevis, D. (2019) *Just Transitions: Social Justice in the Shift Towards a Low-Carbon World*. London: Pluto Press.

Muro, M., Maxim, R. and Whiton, J. (2019) *Automation and Artificial Intelligence: How Machines Are Affecting People and Places*, Metropolitan Policy Program, Washington: Brookings Institution. [Online]. Available: www.brookings.edu/research/automation-and-artificial-intelligence-how-machines-affect-people-and-places/ [Last accessed: 20 May 2020].

Nedelkoska, L. and Quintini, G. (2018) *Automation, Skills Use and Training*, OECD Social, Employment and Migration Working Papers. Paris: OECD. https://doi.org/10.1787/2e2f4eea-en.

Newitz, A. and Stamm, E. (2014) 10 failed utopian cities that influenced the future, *Gizmodo*. [Online]. Available: http://io9.gizmodo.com/10-failed-utopiancities-that-influenced-the-future-1511695279 [Last accessed: 27 April 2020].

Noys, B. (2017) *The City and the City: Accelerationism and Imaging the Urban Future*, Citileaks Academic Lecture, CityLeaks Urban Art Festival, Cologne.

OECD (2019) *Recommendation of the Council on Artificial Intelligence*, OECD/LEGAL/0449. [Online]. Available: www.fsmb.org/siteassets/artificial-intelligence/pdfs/oecd-recommendation-on-ai-en.pdf [Last accessed: 10 May 2020].

Osborne, M. and Frey, C.B. (2018) *Automation and the Future of Work: Understanding the Numbers*, Oxford Martin School. [Online]. Available: www.oxfordmartin.ox.ac.uk/blog/automation-and-the-future-of-work-understanding-the-numbers/ [Last accessed: 8 April 2020].

Pearce, F. (2011) *The Coming Population Crash: And Our Planet's Surprising Future*. Boston: Beacon Press.

Pike, A., MacKinnon, D., Coombes, M., Champion, T., Bradley, D., Cumbers, A., Robson, L. and Wymber, C. (2016) *Unequal Growth: Tackling Declining Cities*. York: Joseph Rowntree Foundation. [Online]. Available: www.jrf.org.uk/sites/default/files/jrf/files-research/tackling_declining_cities_report.pdf [Last accessed: 27 April 2020].

Pitroda, S. and Miailhe, N. (2017) Introduction: The rise of AI & robotics in the city, *Journal of Field Actions*. [Online]. Available: http://journals.openedition.org/factreports/4377 [Last accessed: 22 December 2018].

Rittel, H.W.J. and Webber, M.M. (1973) Dilemmas in a general theory of planning, *Policy Sciences 4* (2): 155–169. https://doi.org/10.1007/BF01405730.

Schleussner, C., Lissner, T.K., Fischer, E.M., Wohland, J., Perrette, M., Golly, A., Rogelj, J., Childers, K., Schewe, J., Frieler, K., Mengel, M., Hare, W. and Schaeffer, M. (2016) Differential climate impacts for policy-relevant limits to global warming: The case of 1.5°C and 2°C, *Earth System Dynamics* 7: 327–351. https://doi.org/10.5194/esd-7-327-2016.

Schwab, K. (2015) The fourth industrial revolution: What it means, how to respond, *Foreign Policy*. [Online]. Available: www.foreignaffairs.com/articles/2015-12-12/fourth-industrial-revolution [Last accessed: 27 April 2020].

Scott, R. (1982) *Blade Runner*, Science Fiction Drama. Burbank, CA: Warner Brothers.

Sennett, R. (2018) *Ethics for the City: Building and Dwelling*. New York: Allen Lane, an imprint of Penguin Books.

Shaviro, S. (2015) Accelerationism without accelerationism, *The Disorder of Things*. [Online]. Available: https://thedisorderofthings.com/2015/11/03/accelerationism-without-accelerationism/ [Last accessed: 8 April 2020].

Smith, C.J., Forster, P.M., Allen, M., Fuglestvedt, J., Millar, R.J., Rogelj, J. and Zickfeld, K. (2019) Current fossil fuel infrastructure does not yet commit us to 1.5°C warming, *Nature Communications 10* (101). https://doi.org/10.1038/s41467-018-07999-w.

Srnicek, N. and Williams, A. (2015) *Inventing the Future: Postcapitalism and a World without Work*. London: Verso Books.

Standing, G. (2011) *The Precariat: The New Dangerous Class*. London and New York: Bloomsbury.

Steele, W. (2019) The city as wild, in Jacobs, K. and Malpass, J. (eds.) *Philosophy and the City*. London: Rowman and Littlefield International.

Taylor, C., Carrigan, J., Noura, H., Ungur, S., van Halder, J. and Dandona, G.S. (2019) Australia's automation opportunity: Reigniting productivity and inclusive income growth, *McKinsey and Company*. [Online]. Available: www.mckinsey.com/featured-insights/future-of-work/australias-automation-opportunity-reigniting-productivity-and-inclusive-income-growth [Last accessed: 27 April 2020].

Tieman, R. (2017) Barcelona: Smart city revolution in progress, *Financial Times*. [Online]. Available: www.ft.com/content/6d2fe2a8-722c-11e7-93ff-99f383b09ff9 [Last accessed: 8 April 2020].

UNESCAP (2018) *Artificial Intelligence in Asia and the Pacific*. Bangkok: UN Economic and Social Commission for Asia and the Pacific. [Online]. Available: www.unescap.org/sites/default/files/ESCAP_Artificial_Intelligence.pdf [Last accessed: 10 May 2020].

Vandenbroeck, P. (2012) *Working with Wicked Problems*. Brussels: King Baudouin Foundation. [Online]. Available: www.kbs-frb.be/en/Virtual-Library/2012/303257 [Last accessed: 10 May 2020].

Verdis, S. and Burdett, R. (2013) Accelerating the pace of city transformations, in Burdett, R., Çavuşoğlu, O. and Verdis, S. (eds.) *City Transformations*. London School of Economics and Political Science. [Online]. Available: https://lsecities.net/wp-content/uploads/2013/10/city-transformations-newspaper_en.pdf [Last accessed: 10 May 2020].

Wells, H.G. (1933) *The Shape of Things to Come: The Ultimate Revolution.* London: Penguin Classics, published in 2005.

Zamyatin, Y. (1921) *We.* Translated by Natasha Randall. London: Vintage Classics, published in 2007.

Zuboff, S. (2019) *The Age of Surveillance Capitalism: The Fight for a Human Future at the New Frontier of Power.* London: Profile Books.

8

BUILDING ETHICAL CITIES

Futures not achieved are only branches of the past: dead branches.

Italo Calvino, Invisible Cities, p. 24

Transformations underway?

A huge and expanding range of actions for change connect responses to the climate crisis with inequality and the need for democratic rights. These range from grassroots movements promoting anti-consumption to policy efforts to internalise environmental externalities. Ethical consumers have long employed voluntary methods in efforts to decarbonise and avoid toxic economic inequalities. Action at the municipal scale lends itself particularly to engendering social welfare and building genuine representation and solidarity with community. Tangential to these movements are assorted business writers and economists advocating greening of capitalism and remedying of social ills via corporate social responsibility and shared value propositions (Elkington 1994; Porter and Kramer 2011). Much lauded, there remain fundamental questions around data and criteria used to measure progress and the contribution that green capitalist projects make in relation to social, sustainability and economic goals (Slaper and Hall 2011). Even some of the leading proponents admit that profit motives (single bottom line) remain paramount and progress has been stymied around the broader aim of promoting systemic, transformative change (Elkington 2018). These contrasting approaches to transforming the capitalist project present a reform/restructure dichotomy; whether to jettison or somehow save capitalism before it wrecks our planet (Mason 2015; Reich 2015; Streeck 2016). To state the obvious, a cursory glance at current trends is enough to conclude that, whatever has been tried to date, hasn't worked (Piketty 2013; Atkinson 2015).

In this chapter we turn our attention to the groundswell of urban-/human- scale transformations targeting the worst excesses of the neoliberal city (i.e. laissez-faire policies, deregulation, rampant poverty and inequality, wage theft, nepotism, cynical privilege, corruption and tokenistic representation). As a starting point, we acknowledge a plethora of sometimes conflicting and often contrasting responses around the idea of the city as a unit of human society that can mobilise and take up the race to transform and decarbonise urban economies at a time when current political ideologies and structures are fraying. What is required is a new political project that respects the power of direct local accountability, rooted in the city as an object and actor. Cities are people, and when we are in active dialogue about building the ethical city, we can create platforms for hopeful, optimistic futures. Concurrently, cities are places where new ways of living within ecological and planetary limits can be imagined. In this chapter, we document examples of actual existing ethical city actions. A shared set of aims provides a basis for co-designing processes that enable groups of residents to co-create the ethical city. This is about the journey, not the destination.

New municipalism

The terms New Municipalism, New Localism, Municipal Socialism and Global Municipalism have received significant attention recently (Roth and Russell 2018; Katz and Nowak 2018; Russell 2019a, 2019b; Centre for Local Economic Strategies 2019a; Ball 2019; Thompson 2020). They share a set of common concerns including community wealth building and the feminisation of politics. The latter is not just about getting more women into politics but includes efforts to insert empathy and care into political action, questioning traditional understanding of strong leadership, seeking to redistribute power within the city and pursuing collective self-governance (Roth and Russell 2018). This is exemplified in fearless cities, a movement that began in Barcelona as an overt network standing up for human rights, radical democracy, feminism and the common good in the face of a world where fear, insecurity, hate, inequality, xenophobia and authoritarianism are used as weapons to divide and undermine communities (Barcelona en Comú et al. 2019). The fearless city describes a progressive urban mindset that is posited against the inaction of national and global capitalist governance and, in doing so, it invariably faces hostility and aggressive push-back from right-wing central governments (Thompson 2020).

Community wealth building, as introduced in Chapter 4, involves re-municipalisation of local assets and services, re-direction of expenditure from anchor institutions (e.g. universities, hospitals, police) towards local suppliers, support for the development of cooperatives and pushing for the adoption of a minimum wage. Universities across the United States have been cooperating in related local neighbourhood revitalisation initiatives since the 1990s (Ehlenz 2018). One example is the Cleveland Model which emerged out of the Greater University Circle Initiative from 2005 onwards (Cleveland Foundation 2013;

Lenham 2014). This was later extended with the launch of the Cleveland Evergreen Initiative – a laundry service, energy business and a food grower's cooperative (Sheffield 2017). The aim was to encourage anchor institutions (including the Cleveland Clinic, Case Western Reserve University, University Hospitals and the Veterans Administration Medical Center) to directly invest in the local community through a range of projects and infrastructure development. The initiative operated in six low-income neighbourhoods suffering from high unemployment, declining local economies and low educational opportunities, and it added around 4,500 local jobs over a five-year period as local institutions used their spending and employment powers to invest and support adjoining neighbourhoods. Similar local efforts across the United States have been promoted as a Metropolitan Revolution, presenting cities as engines of economic prosperity, political renewal and social transformation through innovative interventions from minimum wages to pro-affordable housing policies (Katz and Bradley 2013; Katz and Nowak 2018).

In Baltimore, community wealth building has been applied in the creative arts and entertainment sector in an attempt to revitalise neighbourhoods and enhance their social and economic value (Rich and Tsitsos 2018). Rather predictably, this raises the spectre of gentrification and the role of real estate markets (Smith 2008). Everyone has a right to a walkable, high-service, attractive urban neighbourhood, but inevitably there are powerful private land interests at play, and these ensure that the best spots go to the highest bidder, and as urban quality improves, perhaps through public funds and efforts, so private speculators move in. Clearly, the pros and cons of such initiatives and their attempts to counter regressive property transactions need constant attention (Gaffikin and Perry 2012). Notwithstanding, there is considerable merit in the engagement of anchor institutions with urban revitalisation, especially where a strong and well-understood ethical city framework ensures relationships are based on solidarity rather than exclusion.

A range of local economies initiatives have also spread across the UK in recent years. The Northern Powerhouses Initiative (Martin et al. 2015) was proposed to revive cities in the north of England facing spiralling high street shop vacancy rates. Burgeoning e-commerce sparked an estimated 27 percent shrinkage in town centre activity in 2015–2020 period alone (Moses et al. 2016). The £1.6 billion Stronger Town Fund was conceived to assist poorer towns with a high proportion of residents on low incomes and skill attainment. These centrally coordinated structural funds tend to support Local Enterprise Partnerships (voluntary partnerships between local authorities and businesses) rather than local governments directly (Cox 2019). In contrast, community wealth building is being pursued by local authorities. For example, the Preston Model emerged from the failure of a planned £700 million shopping centre development (Centre for Local Economic Strategies 2015, 2017, 2019b; Brown et al. 2019). Local leaders realised that traditional urban growth models based on attracting inward investment from big infrastructure projects could no longer be relied upon (Chakrabortty 2019; Singer 2016). In 2013, the Council sought ways to apply community wealth building ideas, identifying a collective spend of close to £1 billion from six major anchor institutions. The aim was to redirect their spending, while at the same time enhancing the local economy's capacity to supply goods and services by creating a tightly integrated ecosystem of cooperative enterprises similar to Spain's Mondragon Corporation (Reynolds 2017). By 2019, £74 million was being redirected back into the local economy, with unemployment dropping from 6.5 percent in 2014 to 3.1 percent in 2017 (Centre for Local Economic Strategies and Preston City Council 2019). Preston was named the Most Improved City in the UK in 2018 (PricewaterhouseCoopers 2018).

Building on this experience are planned measures to encourage development of the cooperative sector and ideas for the creation of a community bank. Matthew Brown, Leader of the Council, ties such local economic in initiatives to the emergence of more democratic, worker-owned businesses benefitting from public sector procurement spending (pers.com, March 2020). The aim is enhanced local resilience predicated on the idea that employee-owned firms are less likely to make decisions that alienate employees. Preston's approach, according to the Demos think tank, represents a reimagination, born through adversity, of what is possible in terms of local economic stewardship (Lockey and Glover 2019). It focuses on how local government can use advocacy and spending power to stimulate fairer work and social initiatives, reduce inequality and increase youth employment. In doing so, these local governments recognise the urban politics of proximity, where there is the potential for direct rapport between people and their elected representatives (Russell 2019b; Katz and Bradley 2013). In this context, re-taking control is about cities setting agendas, directing resources and having greater influence over their own future. They can do this most effectively by working with, and learning from, other cities. In this, there is also a tacit acceptance that cities cannot do all that is required on their own; national and global policy settings are imbricated with local policies and opportunities. Mutual cooperation across cities both empowers them in national lobbying and provides practical solidarity in common struggles and experiments. New municipalism therefore is about fighting back against the gradual extraction of wealth and power from the city by regions and nations. It is a demand for the restoration of rights for communities, towns and cities that have been eroded over time (Centre for Local Economic Strategies 2019b).

Thompson (2020, p. 3) identified a tripartite typology for the conceptualising of New Municipalism. As shown in Table 8.1, these are: (1) *Platform Municipalism* – exemplified by Barcelona – working against state and platform capitalism (businesses that provide hardware/software foundations for others to operate on) via civil society mobilisation and the use of digital technologies; (2) *Managed Municipalism* – in the form of the Cleveland and Preston models – aiming to reinvigorate the local state by democratising the urban economy through technocratic engineering; (3) *Autonomist Municipalism* – aiming for a stateless polis of confederated cooperatives, communities and assemblies through collective

TABLE 8.1 Three types of new municipalism

	Platform Municipalism	Managed Municipalism	Autonomist Municipalism
Examples	Barcelona (Spain)	Preston (UK), Cleveland (USA)	Rojava, Jackson (USA)
Key Characteristics	Social movement-driven, rooted in urban politics of inhabitance/proximity	Technocratic/think tank project, rooted in community wealth building	Social movement-driven, rooted in place-based cultural/racial identity
Catalysts	Financialisation, dispossession, neoliberal austerity	Neoliberal failure to resolve economic decline, urban shrinkage, deindustrialisation	Racist democratic, eco-destructive practices of colonial capitalist state
Strategy/Aims	Transform local state through dual power (in, against and beyond the state)	Reclaim/regenerate local economy (retooling the state from the inside)	Realise democratic eco-socialist self-governance (building a new polis outside)
Institutional Forms	New state institutions (digital platforms, co-ops, participatory budgeting, popular assemblies)	Community-owned local institutions (worker-owned co-ops, community land trusts, anchor institutions	Confederation of autonomous self-governing communes and co-ops
Historical Influences	Anarcho-syndicalism, federalism	Municipal socialism, Fabianism (UK), guild socialism	Anarchism, national self-determination struggles
Theoretical Influences	Feminism, Right to the City (Lefebvre), commons (Federici), libertarian municipalism (Bookchin)	Cooperatives (Mondragon), pluralist commonwealth (Alperovitz)	Feminism, degrowth, eco-socialism, libertarian municipalism, communalism
Spatial Imaginaries	Urban platforms, confluences (tides, overflow), the urban everyday	Leaky bucket, containing trickle-out economics, anchor institutional flows	Confederated communes, bioregionalism
Economic Interventions	Socialisation of platform capitalism (technological sovereignty, platform cooperatives)	Localised supply chains (progressive procurement, worker-owned cooperatives)	Non-commodified circuits of value (social reproduction, communing)

Source: Thompson (2020, pp. 11, 12).

self-organising, such as Jackson, Mississippi, focusing on developing a solidarity economy, ecological self-sufficiency and non-monetary exchange. (ibid., 13).

While each form of municipalism is inspiring in one way or another, the future realities for these nascent modes of local activism are stark without broader application. Nation states invariably control taxation and key societal and economic functions, and there are limits to their tolerance of the wayward polis. Whether these campaigns are directed towards economic emancipation or political autonomy, threats to central power are invariably met with predictable force. Moreover, disruptions provide incumbents with further opportunities to consolidate and reinforce current power relations, whether these are through taking advantage of technological change and digital disruption, concentration of power in transnational oligarchy corporations, economic stagnation in some places and uncontrolled growth in others or perverse patterns of inequality and wealth distribution. Similarly, pandemics, extreme weather events arising from climate change and other so-called external threats provide opportunities for incumbent governments to remove existing rights; just as they also provide openings for organised resistance at the local scale.

Scale is important and there are good reasons why some services and infrastructure are coordinated regionally, rather than at city scale. The local is potentially a trap (Purcell 2006). City agency can be repurposed regressively, paralleling and exacerbating, wider fragmented, individualist, tribal, isolationist, nationalist societal tendencies rather than diverging from and ameliorating. Ultimately, even where progressive anchor institutions work to preserve economic vitality and generate community wealth, localities may not be immune to, and could be undermined by, the push to go online, cashless and borderless with concomitant cost-savings and efficiencies (Mitchell and LaVecchia 2018).

All the same, cities are far from powerless when engaging in ambitious and far-reaching social urban experiments. There are opportunities to test a wide range of innovations designed to provoke shifts in the patterns of power distribution within the city (Russell 2019a). As cities innovate and experiment, they have to deal with the fact that 'every innovation suffers by definition from a mismatch between the ways people currently do things and the ways they might do them' (Sennett 2018, p. 14). This implies innovations will always need to overcome vested interests. Engaging in radical social-cultural transformations, which put collectivist approaches at the centre, may require that proponents of the ethical city seek to supersede old rule books and work around incumbent power structures. As mentioned previously, our central contention is that the adoption of ethical urban development approaches requires embodiment of principles of justice, empathy, care, human rights and responsibility. We reiterate here that that this could create new affordances for alternative pathways to emerge, such as prosperity without growth, sharing economies, collaborative consumption and even degrowth strategies (Jackson 2009; Botsman and Rogers 2011; McLaren and Agyeman 2015; Sharp 2016; Gleeson and Alexander 2018; Nelson 2018).

While recognising the significance of local innovation, a key challenge facing New Municipalism is for localities to develop a variegated scalar strategy of transformation that retains the democratic essence that underpins them without falling into the trap of a particular localism (Russell 2019a; Purcell 2006). This is about the quest for a society that is 'more equal and more democratic than most today, with wealth and power spread through the entire social body rather than hoarded at the top' (Sennett 2018, p. 8).

Inclusive local economies

Minamata, a small town in Kumamoto Prefecture, Japan, is perhaps best known as the site of a tragic mercury pollution incident that came to light in the 1950s resulting in 1,784 deaths. In recent years, however, this small town with a population of just over 25,000 has been struggling economically. In response, the municipal government undertook a health check of the local economy that involved an analysis of money circulation patterns, also known as a regional economic cycle analysis. This involved an evaluation of local production of goods and services, distribution of income from labour and capital as well as spending in terms of consumption and investment including that of financial institutions. The aim was to identify financial flows into and out of the locality. The study revealed that most residents spent a large part of their income outside the local economy (roughly half of all shopping occurred outside), only 20–30 percent of funds managed by local financial institutions were reinvested in the town and 8 percent of the total regional production was spent to import energy. The size of the local gross production for Minamata in 2010 was JPY 108.8 billion, or around £836 million, implying that close to 7.8 percent (£66 million) was leaving the locality each year just in energy costs alone (Edahiro 2016).

This mirrors the experience of many local economies where a significant proportion of regional domestic production (or gross value added – GVA) leaks out. A study in Chicago revealed that for every $100 spent with local businesses, 68 percent remained in the local economy, while for a corporate chain store the amount was closer to 43 percent (Cunningham 2004). For an online purchase the equivalent percentage would be close to zero. One response has been the introduction of local currencies such as the Bristol Pound (£B) in an attempt to encourage people to shop at independent local businesses. Its managing director, Diana Finch also noted the currency's role in engendering local pride (pers. com., March 2020). As Finch explained, following the GFC a lot of thought went into how best to insulate Bristol from global economic shocks, by keeping more money in the local economy. The Bristol Pound grew rapidly, with beautifully designed notes and digital money supported by a phone app, to become the UK's largest local currency with over 1,500 individual members and 500 or more business members accepting the currency. It apparently reinforced a local sense of community (Ferreira and Perry 2014) and promoted values in support of a 'local and circular economy which is deemed to be more sustainable, ethical and resilient in the face of corporate capitalism and rampant globalisation' (Johnson and Harvey-Wilson 2018, p. 29). By 2019, there was an estimated £B 1 million in circulation, mainly in the form of e-money, with only 10 percent in paper money form.

However, the £B has had limited impact on localisation of procurement spending and the promotion of local production of goods and services (Marshall and O'Neill 2018). The £B cannot effectively address the mobility of capital, the power of global corporations and the expansionary logic of capitalism. More-over, in the rush to a cashless society, current £B transaction charges are insuf-ficient to cover operating costs, and the declining popularity suggests a new model is required with a clearer value proposition for local businesses. Elsewhere, local currencies have also been criticised for failing to address the needs of lower socio-economic groups in the community and it is suggested that they work best in areas with strong, liberal, middle-class communities (Harford 2008). In cities in the US Rustbelt, where local currencies are perhaps most needed, they have not taken root (Collom 2005). The main point here is that we should not mistakenly view local currencies as a silver bullet but instead recognise their transformative potential as part of a broader package of measures.

Another measure receiving increased attention both locally and nationally is universal basic income (UBI). There is a long history of experimentation with UBI and an emergent appreciation that it could play a significant role in addressing the needs of socio-economic groups impacted by a range of disrup-tive forces including AI, automation and the pandemic (Standing 2017; Bregman 2017; Livni 2020). While there are various UBI models, essentially every citizen regularly receives a guaranteed unconditional sum of money. A study by the UK Royal Society for the Encouragement of Arts, Manufactures and Commerce (RSA) suggested that UBI supports people in nurturing their lives and frees them to create a new future (Painter and Thoung 2015). In Finland, a two-year UBI experiment was implemented by the national government in 2017–2018 with preliminary research findings revealing mixed results but indicating that recipients (compared to a control group of unemployed individuals) suffered sig-nificantly fewer problems related to health, well-being and stress. They were also more positive about the future and more trusting in other people (Kangas et al. 2019). A subsequent report revealed that:

> basic income recipients experienced less stress and symptoms of depression and better cognitive functioning than the control group. In addition, the financial well-being of basic income recipients was better. They reported to be more often able to pay their bills on time.
>
> *(ibid. 2020, p. 188)*

A three-year experiment was initiated at the local level in Ontario, Canada in 2018 providing a basic income to 4,000 residents (with a comparison group of another 2,000 residents) in three communities in Hamilton, Thunder Bay and

Lindsey. The participation criteria targeted individuals aged between 18 and 64 living on a low income of less than CAD$34,000 per year (Ontario Government 2019). Data from the baseline survey of participants (at the start of the experiment) revealed 35 percent in employment and 9.8 percent in full-time education (Blueprint 2018). In addition, 43.6 percent of participants stated that they had real financial problems in the past 12 months, with another 50.3 percent indicating that they sometimes struggle to keep up. Perhaps of greatest concern was the fact that 36.5 percent were suffering severe psychological distress and another 44.4 percent were dealing with moderate psychological distress. Further, 51.2 percent specified that they had to cope with severe food insecurity and 5.4 percent were either homeless or living in an institution or collective dwelling. Unfortunately, following local elections the new conservative government announced that the pilot would end on 31 March 2019, resulting in a class action lawsuit by some participants against the Ontario Government for failure to uphold a contractual commitment (Bizarro 2018; Mulvale 2018).

The Barcelona Minimum-Income (B-MINCOME) project is another UBI experiment aimed at combating poverty and inequality. This partnership, between the city of Barcelona and local research institutions, was funded by the European Union via the Urban Innovative Actions Lab. It involved a three-year, €17 million investment in Besos, the most deprived region in the city, involving 2,000 households (half of which functioned as a control group) receiving between €100 to €1676 euros per month in the form of Municipal Inclusion Support (MIS), depending on their circumstances. Interventions ranged from economic aid complemented by training programmes, cooperative economic activities, housing subsidies and community participation. Expected impacts included enhanced labour market participation, food security, housing security, energy access, education participation and attainment (McFarland 2017). Recipients of MIS were required to ensure that this funding was used for only basic needs (food, education, clothing and housing) and agreed to provide documentary evidence of their expenditure or risk disqualification. The project also used a new digital currency (referred to as a social currency or real economy currency) with 25 percent of the MIS being paid via this method. The B-MINCOME project reached completion in December 2019, and a preliminary evaluation report (covering the first year of implementation) identified improvements in terms of reducing the rates of food insecurity and material deprivation. However, there was no significant progress in reducing housing insecurity and there were no changes in terms of the self-perceived health. In aggregate across the ten participant neighbourhoods, general satisfaction with life increased by 27 percent compared to the baseline survey as shown in Table 8.2. This is based on the methodology used by the European Social Survey which found the average level of satisfaction with life across Spain in 2016 was 7.46 points out of 10. Similar figures for the Eix Besos community in Barcelona were 5.04 in 2017 on commencement of the experiment and rising to 6.45 in 2018 (Barcelona City Council 2019).

TABLE 8.2 Changes in levels of general satisfaction with life according to B-MINCOME participation type: 0 = completely unsatisfied, 10 = completely satisfied

Participation Type or Group	Responses	Average Level of General Satisfaction		
		Survey 1 – Sept. 2017	Survey 2 – Nov. 2018	Variation
Unlimited SMI	174	4.80	6.39	33.1%
Limited SMI	128	5.13	6.29	22.7%
SMI + Community participation (Unlimited)	99	5.62	6.73	19.8%
SMI + Community participation (Limited)	94	4.85	6.09	25.4%
SMI + Training and Employment (Non-conditional)	52	5.27	7.02	33.2%
SMI + Training and Employment (Conditional	43	4.77	7.07	48.3%
SMI + Social entrepreneurship (Non-conditional)	34	5.09	6.35	24.9%
SMI + Social entrepreneurship (Conditional)	37	5.03	6.30	24.9%
SMI + Room rental aid (Non-conditional)	5	4.80	7.00	45.8%
SMI + Room rental aid (Conditional)	3	3.00	2.67	-11.1%
Total	669	5.04	6.45	27.9%

Source: Barcelona City Council (2019, p. 27).

There are also numerous examples in the Global South, including the Citizen's Basic Income Initiative in Maricá, Brazil (population of 157,000). Under this programme more than 42,000 residents received the equivalent of 130 Brazilian Real (US$33) per month with the result that most were lifted above the national poverty line (Bendix 2019). Unlike other UBI schemes, the Maricá case has had a dedicated funding stream tied into royalties from nearby Petrobras oil fields (Matthews 2019). A local digital social currency (Moeda Mumbuca) has formed part of what is described as a solidarity economy including social benefits, city salaries and payments to registered local traders. Backed by a community bank (Banco Mumbuca), the currency has been funded through the municipal budget with registered residents receiving the Mumbuca debit/credit card (Katz and Ferreira 2020).

While the approach and political circumstances surrounding the UBI pilots vary considerably, they illustrate the potential of these social experiments devised to address pressing issues of inequality, poverty and vulnerability. A complementary measure that can be utilised by local authorities to promote enhanced community engagement in an inclusive local economy is participatory budgeting.

In Chapter 4, we briefly introduced the experience of Porte Alegre, Brazil. Another much lauded case from elsewhere in Latin America is that of Medellín, Colombia which adopted participatory budgeting in 2004, codified in a Municipal Council Agreement in 2007. The goal was to give people more say in how the city spends its funds and to restore public trust in local institutions (McLaren and Agyeman 2015).

Decarbonising local economies

Given these experiences with local economic innovation, how might local urban economies transform themselves in the face of the climate crisis? As discussed in Chapter 7, urban responses to the woeful lack of leadership by nation states and global institutions on climate change have gathered significant momentum, prompting multiple networks, experiments and a plethora of concerned philanthropists and grassroots campaigns. Across these initiatives, a critical ethical question to be addressed is how cities can achieve the goal of an inclusive, prosperous local economy while at the same time ensuring the attainment of environmental sustainability, including, specifically, decarbonising.

In this context, there has been a recent shift in understanding of deep decarbonisation from a focus on purely technical transitions towards a requirement for broader societal change. For example, the 2019 Carbon-Roadmap for the City of Leeds argues that delivering further changes needed to meet ambitious targets – especially in the coming decade when fast and deep carbon cuts are required – will depend on transformative action in all parts of the city. The Roadmap submits that a key challenge is to ensure that the transition is a just and inclusive one – with steps being taken to ensure that people and places are not left behind and that all social groups and economic sectors participate in and benefit from the transition (Leeds Climate Commission 2019). Changes to the local economic structure have meant that Leeds (population of 474,000) witnessed an emissions decline from 6.8 million to 3.95 million tonnes CO_2 emission between 2005 and 2018. The goal is to reduce emissions by 85 percent in 2030 and to reach zero carbon by 2050. As shown in Figure 8.1, the approach adopted by Leeds involves economically viable, technically viable, innovative and behavioural actions. Under economically viable actions it is estimated, for instance, the city could save £277 million per year in total energy bills. With regard to behavioural actions, the Roadmap recognises that the city's carbon footprint includes emissions that occur outside the urban boundary as a result of demands for goods and services that are consumed within the city. These so-called embedded or consumption-based emissions can only be addressed by changes in the patterns of consumption (as discussed in Chapter 7).

Alongside the rush to make commitments to deep emissions cuts, many cities are seeking to integrate carbon emission reduction strategies with spatial planning and responses to pressing social challenges. The City of Berlin, for example, aims to become carbon neutral by 2050. In 2014, the city commissioned a feasibility

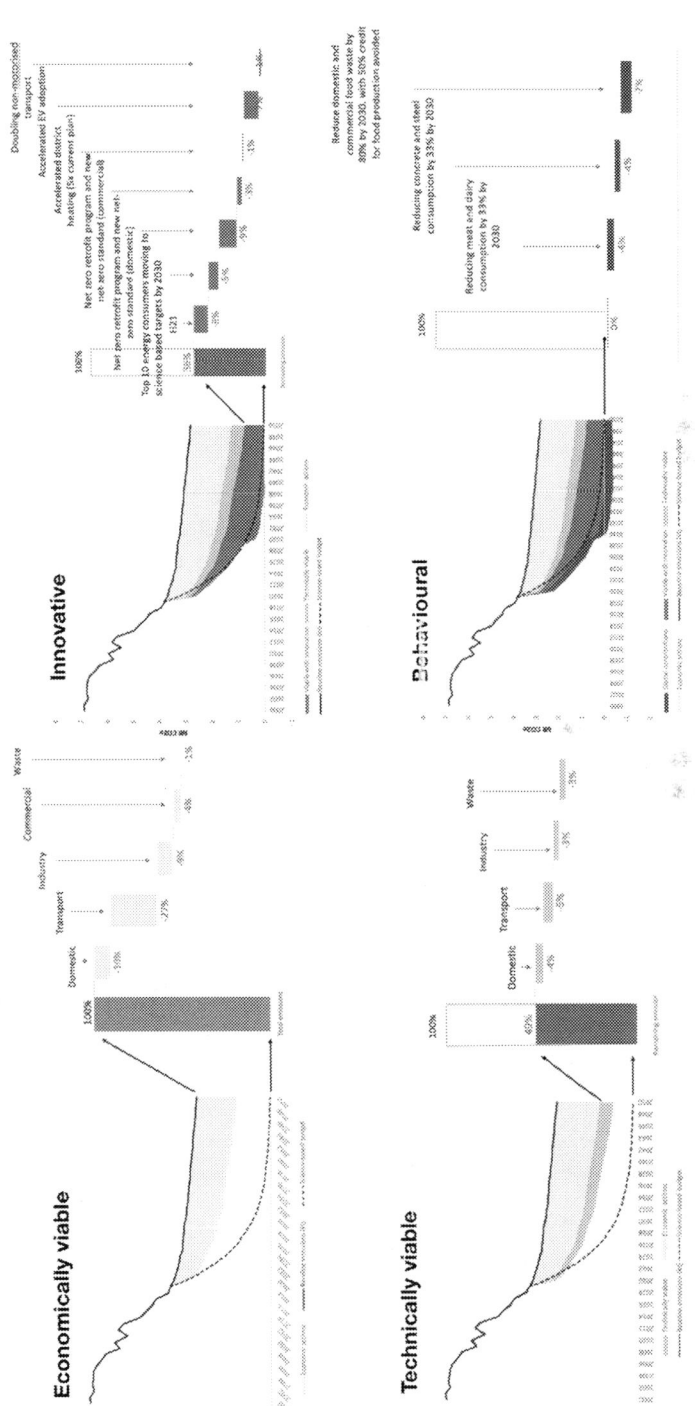

FIGURE 8.1 Viable emission reduction actions under the Leeds Carbon-Roadmap

Source: Leeds Climate Commission 2019.

study from the Potsdam Institute for Climate Impact Research revealing that while emissions had declined by 27 percent between 1990 and 2010, they still stood at 21.3 million tonnes and needed to decline by 2 percent per annum in order to reach the carbon neutral target of 4.4 million tonnes CO_2 emission in 2050 (about 85 percent less than 1990 levels) as shown in Figure 8.2 (Reusswig et al. 2014). This feasibility study pre-dates the 2018 IPCC special report and is based on the assumption that the goal should be to keep global warming below the dangerous 2°C threshold. Nevertheless, even the identified emission reductions would require a new approach to urban development as proposed in the Berlin 2030 – Urban Development Concept (Berlin City Council 2015). The feasibility of achieving the necessary deep emission reductions requires direct engagement with everyone in the city. Consequently, Berlin partnered with the Potsdam Institute once more in 2017 to implement the Climate Neutral Berlin Living Lab. This involved 100 households in the city testing what climate action means in everyday life using a carbon tracker and documenting measures taken to improve their individual climate footprint over a one-year period. Intermediate results showed that the participants were able to reduce their footprints from 11.6 tonnes CO_2 emission per person down to 7.3 tonnes CO_2 emission – a 37 percent reduction.

Oslo City Council has been promoting an integrated approach through its 2016 Climate and Energy Strategy, initially aiming to cut emissions in half by 2020 and by 95 percent in 2030 compared to 1990 (City of Oslo 2016). Initiatives include Økern Centre Living Lab, focused on a development of eco-dwellings, and a Climate Budget fully integrated with the municipal financial system and including funded emission reduction measures (City of Oslo 2018). The 2020 target was revised due to delays with a major carbon capture and sequestration project at Klemestrud. However, the point here is not so much about the missing of targets or the achievement of budgets; the benefits of the transparent Carbon Budget is that it ensures annual measures are clearly visible to residents and others. They can also see what and where additional work and investment is

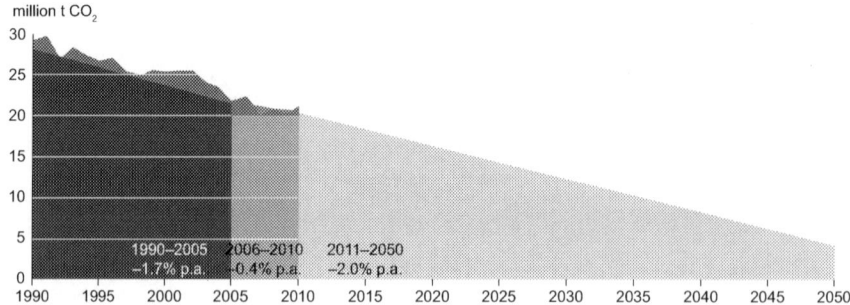

FIGURE 8.2 Reduction of CO_2 emissions in Berlin 1990–2010 and the scenario until 2050

Source: Reusswig et al. (2014, p. 8).

proposed in order to bring about the required emission reductions; for example, by 2024 all taxis operating in Oslo are required to be 'zero GHG emission'.

Resurgent attempts at integrated city planning include One New York: The Plan for a Strong and Just City (New York City 2015). This introduced a range of measures to alleviate urban poverty (including a minimum wage) so as to lift 800,000 New Yorkers out of poverty by 2025, while also reducing CO_2 emissions by 80 percent by 2050 (compared to 2005 levels). It also commits to convert all energy to 'clean' sources by 2040 (latest data indicates 27 percent). Progress with plan implementation is tracked against 55 quantitative indicators annually. The 2019 update included nine thematic volumes around local democracy, inclusive economy, thriving neighbourhoods, healthy lives, education, liveable climate, mobility and modern infrastructure (New York City 2019). Interim achievements included a US$15-per-hour minimum wage, delivery of 275,000 affordable homes and a reduction in greenhouse gas emissions of 17 percent compared to 2005. In the absence of federal leadership these local decarbonisation initiatives are linked to the Paris Agreement and the 1.5°C target (New York City 2017). The city has also sought to link to the global movement on local carbon reporting and was the first city to report to the United Nations in July 2018 on its efforts to comply with the sustainable development goals through the submission of a voluntary local review (New York City 2018).

Bristol council has also sought increased local decarbonisation efforts. A 'mini-Stern review' for the city (Gouldson and Millward-Hopkins 2015) proposed that by investing just 0.4 percent of GVA each year for ten years, it would be possible to create 2,000 jobs in the low-carbon goods and services sector, save around £220 million in energy bills and reduce CO_2 emissions by 33 percent by 2025 compared to 2005. The report's authors argue that this transition:

> depends on political and social capital as well as financial capital. The levels of ambition, investment and activity needed to exploit the available potential are very significant indeed. Enormous levels of investment are required, and major new initiatives are needed with widespread and sustained influence.
>
> *(ibid., p. 44)*

These and other recommendations are incorporated in the Bristol One City Plan (Bristol City Council 2020) aimed to support the emergence of an inclusive, sustainable city that both breaks down social fractures and inequalities and reaches carbon neutrality. The goal is to ensure that by 2050 the city is carbon neutral and 95 percent of all energy consumed is generated from clean sources. Bristol Energy has been established as a not-for-profit municipally owned energy company to drive decarbonisation efforts including renewable energy-related infrastructure, innovative energy supply and demand management solutions. It also promotes the growth of electric vehicles, charging infrastructure and autonomous vehicles (Bristol City Council 2018). In parallel, plan implementation engagement mechanisms include a biannual city gathering, city leadership forums that bring together representatives from key organisations, thematic boards and monthly city office drop-in meetings (i.e. open to the public).

Reimagining cities as ethical

All examples presented in this chapter (community wealth building, local currencies, UBI, participatory budgeting, rapid decarbonisation of local economies, One Plans) go beyond emphasis on deontological ethics (i.e. ethical codes of practice, ethics commissions etc.) and into the realm of teleological ethics – where engagement between local government and community is motivated by specified outcomes and consequences. Big questions are necessarily being asked – How do we eliminate poverty?; How do we ensure everyone can make a meaningful contribution to society?; How do we create an economy that works for all?; Radical solutions are being tried and tested: community wealth building to revive the local economy; local currencies to support small and medium sized businesses; universal basic income to engender equity, health and well-being etc. Through experimentation, some local communities are pursuing preference utilitarianism – searching for solutions that fulfil the greatest amount of personal and collective interests, as opposed to actions that generate the greatest amount of individual pleasure.

City ethical innovations overtly linked to the political project of transformation offer genuine albeit unpredictable prospects. The diversity of ongoing innovations around new municipalism, the inclusive economy and local decarbonisation is laudable. The examples presented in this chapter are predominantly from so-called post-industrial economies. This does not imply a lack of innovative examples in cities across all global regions. Rather, there is a particular urgency in the search for new modes of collaboration and cooperation within and across communities in declining and fragmented cities in the rustbelt of the United States and the former industrial areas of the UK. Other, perhaps even greater, urgencies apply to rapidly growing cities with burgeoning informal settlements and accelerating poverty. The examples we have shared in this chapter highlight the need for ethical cities leadership characterised by recallability, accountability and a singular focus upon tackling inequality and climate change in an integrated way.

The practicality of building ethical cities from the ground up implies interconnected issues ranging from openness and anti-corruption to fairness – both within the city and across cities, generations and species. It also inevitably demands creativity and imagination in proposing new, innovative solutions to problems. So far, we have outlined what is only the tip of the iceberg of ideas and their ethical orientation. In any event cities (people, governments, places) cannot just keep doing the same old things and somehow expect to get different and better results. 'They' can instead advance alternative approaches and ideas that work to reshape urban agendas in more ethical directions. The ethical city, with appropriate virtues and sentiments, can exist and be built upon – from the ground up.

References

Atkinson, A.B. (2015) *Inequality: What Can Be Done?* Cambridge: Harvard University Press.

Ball, J. (2019) What is 'New Municipalism', and can it really combat austerity? *New Statesman.* [Online]. Available: www.newstatesman.com/spotlight/2019/04/what-new-municipalism-and-can-it-really-combat-austerity [Last accessed: 17 March 2020].

Barcelona City Council (2019) *Report on the Preliminary Results of the B-MINCOME Project (2017–18): Combining a Guaranteed Minimum Income and Active Social Policies in Deprived Urban Areas of Barcelona.* [Online]. Available: https://ajuntament.barcelona. cat/dretssocials/sites/default/files/arxius-documents/results_bmincome_eng.pdf [Last accessed: 18 March 2020].

Barcelona en Comú, Bookchin, D. and Colau, A. (2019) *Fearless Cities: A Guide to the Global Municipalist Movement.* Oxford: New Internationalist.

Bendix, A. (2019) A Brazilian city is giving a third of its residents $33 per month: Part of one of the largest basic income programs in the world, *Business Inside.* [Online]. Available: www.businessinsider.com/marica-brazil-basic-income-policy-monthly-payments-2019-11 [Last accessed: 15 April 2020].

Berlin City Council (2015) *Berlin Strategy: Urban Development Concept Berlin 2030, State Department for Urban Development and the Environment.* [Online]. Available: www. stadtentwicklung.berlin.de/planen/stadtentwicklungskonzept/download/strategie/ BerlinStrategie_Broschuere_en.pdf [Last accessed: 18 March 2020].

Bizarro, S. (2018) Ontario, Canada: Reactions to Ontario basic income pilot cancelation, *Basic Income Earth Network.* [Online]. Available: https://basicincome.org/news/2018/09/ ontario-canada-reactions-to-ontario-basic-income-pilot-cancelation/ [Last accessed: 18 March 2020].

Blueprint (2018) *Ontario Basic Income Pilot Baseline Survey: Preliminary Analysis.* [Online]. Available: https://hamiltonpoverty.ca/preview/wp-content/uploads/2019/01/OBIP-Baseline-Survey-Preliminary-Analysis-1.pdf [Last accessed: 18 March 2020].

Botsman, R. and Rogers, R. (2011) *What's Mine Is Yours: How Collaborative Consumption Is Changing the Way We Live.* New York: Collins.

Bregman, R. (2017) *Utopia for Realists and How We Can Get There.* London and New York: Bloomsbury.

Bristol City Council (2018) *City Leap Prospectus: A Soft Market Testing Exercise for a Potential Partners Exercise in a City Scale Low Carbon, Smart Energy Infrastructure Programme.* [Online]. Available: www.energyservicebristol.co.uk/wp-content/pdf/ City_Leap_Prospectus%204-5-18.pdf [Last accessed: 11 May 2020].

Bristol City Council (2020) *One City Plan 2020: A Plan for Bristol to 2050.* [Online]. Available: www.bristolonecity.com/wp-content/uploads/2020/01/One-City-Plan_2020. pdf [Last accessed: 11 May 2020].

Brown, G., Burch, D. and Todd, M. (2019) *Community Business and Anchor Institutions.* Manchester: CLES. [Online]. Available: https://cles.org.uk/wp-content/uploads/2019/02/ Community-business-and-anchor-institutions-Digital.pdf [Last accessed: 11 May 2020].

Calvino, I. (1972) *Invisible Cities.* London: Vintage Books, published in 1997.

Center for Local Economic Strategies (2015) *Creating a Good Local Economy: The Role of Anchor Institutions*. Manchester: CLES. [Online]. Available: https://cles.org.uk/wp-content/uploads/2016/10/Anchor-institutions.pdf [Last accessed: 11 May 2020].

Centre for Local Economic Strategies (2017) *Community Wealth Building through Anchor Institutions*. Manchester: CLES. [Online]. Available: https://cles.org.uk/wp-content/uploads/2017/02/Community-Wealth-Building-through-Anchor-Institutions_01_02_17.pdf [Last accessed: 11 May 2020].

Centre for Local Economic Strategies (2019a) *New Municipalism in London*. Manchester: CLES. [Online]. Available: https://cles.org.uk/wp-content/uploads/2019/04/New-Municipalism-in-London_April-2019.pdf [Last accessed: 11 May 2020].

Centre for Local Economic Strategies (2019b) *Community Wealth Building 2019: Theory, Practice and Next Steps*. Manchester: CLES. [Online]. Available: https://cles.org.uk/wp-content/uploads/2019/09/CWB2019FINAL-web.pdf [Last accessed: 11 May 2020].

Centre for Local Economic Strategies and Preston City Council (2019) *How We Built Community Wealth in Preston: Achievements and Lessons*. Manchester: CLES. [Online]. Available: https://cles.org.uk/wp-content/uploads/2019/07/CLES_Preston-Document_WEB-AW.pdf [Last accessed: 11 May 2020].

Chakrabortty, A. (2019) In an era of brutal cuts, one ordinary place has the imagination to fight back, *The Guardian*. [Online]. Available: www.theguardian.com/commentisfree/2019/mar/06/brutal-cuts-fight-back-preston-dragons-den [Last accessed: 11 May 2020].

City of Oslo (2016) *Climate and Energy Strategy*. [Online]. Available: www.klimaoslo.no/wp-content/uploads/sites/88/2018/06/Climate-and-Energy-Strategy-2016-English.pdf [Last accessed: 18 March 2020].

City of Oslo (2018) *Climate Budget 2018*. [Online]. Available: www.klimaoslo.no/wp-content/uploads/sites/88/2018/02/Climate-Budget-English.pdf [Last accessed: 18 March 2020].

Cleveland Foundation (2013) *Cleveland's Greater University Circle Initiative: A Partnership between Philanthropy, Anchor Institutions and the Public Sector*. [Online]. Available: https://community-wealth.org/sites/clone.community-wealth.org/files/downloads/Cleveland%27s%20Greater%20University%20Circle%20Anchor%20Initiative.%20Case%20Study.pdf [Last accessed: 1 May 2020].

Collom, E. (2005) Community currency in the United States: The social environments in which it emerges and survives, *Environment and Planning A 37*: 1565–1587. https://doi.org/10.1068/a37172.

Cox, E. (2019) *What Will be the Impact of the Stronger Towns Fund?* Royal Society for the encouragement of Arts, Manufactures and Commerce. [Online]. Available: www.thersa.org/discover/publications-and-articles/rsa-blogs/2019/03/stronger-towns-fund [Last accessed: 11 May 2020].

Cunningham, M. (2004) *The Andersonville study of retail economics*. Chicago, IL: Civil Economics. [Online]. Available: https://community-wealth.org/sites/clone.community-wealth.org/files/downloads/report-civic-economics.pdf [Last accessed: 22 April 2020].

Edahiro, J. (2016) Community building through local money circulation analysis: The case of Minamata, *Japan for Sustainability*. [Online]. Available: www.japanfs.org/en/news/archives/news_id035639.html [Last accessed: 23 November 2019].

Ehlenz, M.M. (2018) Defining university anchor institution strategies: Comparing theory to practice, *Planning Theory and Practice 19* (1): 74–92. https://doi.org/10.1080/14649357.2017.1406980.

Elkington, J. (1994) Towards the sustainable corporation: Win-win-win business strategies for sustainable development, *California Management Review 36* (2): 90–100. https://doi.org/10.2307/41165746.

Elkington, J. (2018) 25 years ago I coined the phrase 'triple bottom line': Here's why it's time for a rethink, *Harvard Business Review*. [Online]. Available: https://hbr.org/2018/06/25-years-ago-i-coined-the-phrase-triple-bottom-line-heres-why-im-giving-up-on-it [Last accessed: 10 May 2020].

Ferreira, J. and Perry, M. (2014) *Research Highlights: The Bristol Pound*, Brunel University. [Online]. Available: https://bristolpound.org/wp-content/uploads/library/Brunel_University_Research_Feedback.pdf [Last accessed: 11 May 2020].

Gaffikin, F. and Perry, D.C. (2012) The contemporary urban condition: Understanding the globalizing city as informal, contested and anchored, *Urban Affairs Review 48* (5): 701–730. https://doi.org/10.1177/1078087412444849.

Gleeson, B. and Alexander, S. (2018) *Degrowth in the Suburbs: A Radical Urban Imaginary*. Basingstoke and New York: Palgrave Macmillan.

Gouldson, A. and Millward-Hopkins, J. (2015) *The Economics of Low Carbon Cities: A Mini-Stern Review of the City of Bristol*, University of Bristol. [Online]. Available: http://bristol.ac.uk/cabot/media/documents/bristol-low-carbon-cities-report.pdf [Last accessed: 11 May 2020].

Harford, T. (2008) It's like money, but with no dead presidents: Do local currencies really help the communities that use them? *Slate*. [Online]. Available: https://slate.com/culture/2008/05/do-local-currencies-really-help-the-communities-that-use-them.html [Last accessed: 24 April 2020].

Jackson, T. (2009) *Prosperity without Growth: Foundations for the Economy of Tomorrow*. London and New York: Routledge.

Johnson, S. and Harvey-Wilson, H. (2018) *A Realist Evaluation of the Bristol Pound*, Centre for Development Studies, University of Bath. [Online]. Available: https://purehost.bath.ac.uk/ws/portalfiles/portal/202188337/bpd56.pdf [Last accessed: 11 May 2020].

Kangas, O., Jauhiainen, S., Simanainen, M. and Ylikanno, M. (eds.) (2019) *The Basic Income Experiment 2017–2018 in Finland: Preliminary Results*, Ministry of Social Affairs and Health, Helsinki. [Online]. Available: http://julkaisut.valtioneuvosto.fi/bitstream/handle/10024/161361/Report_The%20Basic%20Income%20Experiment%2020172018%20in%20Finland.pdf?sequence=1&isAllowed=y [Last accessed: 11 May 2020].

Kangas, O., Jauhiainen, S., Simanainen, M. and Ylikanno, M. (2020) *Evaluation of the Finnish Basic Income Experiment* (in Finnish), Ministry of Social Affairs and Health, Helsinki. [Online]. Available: https://julkaisut.valtioneuvosto.fi/bitstream/handle/10024/162219/STM_2020_15_rap.pdf?sequence=1&isAllowed=y [Last accessed: 11 May 2020].

Katz, B. and Bradley, J. (2013) *The Metropolitan Revolution: How Cities and Metros Are Fixing Our Broken Politics and Fragile Economy*. Washington: Brookings Institution Press.

Katz, B. and Nowak, J. (2018) *The New Localism: How Cities Can Thrive in the Age of Populism*. Washington: Brookings Institution Press.

Katz, P. and Ferreira, L. (2020) What a solidarity economy looks like, *Boston Review*. [Online]. Available: https://bostonreview.net/class-inequality/paul-katz-leandro-ferreria-brazil-basic-income-marica [Last accessed: 15 April 2020].

Leeds Climate Commission (2019) *A Science-Based Carbon Budget, Carbon Targets and Carbon-Roadmap for Leeds*. [Online]. Available: www.leedsclimate.org.uk/sites/default/files/Leeds%20Carbon%20Roadmap%20v4.pdf [Last accessed: 18 March 2020].

Lenham, R. (2014) Rust-belt recovery: The Cleveland model as economic development in an age of economic stagnation and climate change, *Pepperdine Policy Review 7*. [Online]. Available: http://digitalcommons.pepperdine.edu/ppr/vol7/iss1/6 [Last accessed: 11 May 2020].

Livni, E. (2020) Coronavirus sparks support for universal basic income in unlikely places, *Quartz*. [Online]. Available: https://qz.com/1819369/coronavirus-sparks-support-for-universal-basic-income/ [Last accessed: 18 March 2020].

Lockey, A. and Glover, B. (2019) *The Wealth within: The Preston Model and New Municipalism*. London: Demos. [Online]. Available: https://demos.co.uk/wp-content/uploads/2019/06/June-Final-Web.pdf [Last accessed: 11 May 2020].

Marshall, A.P. and O'Neill, D.W. (2018) The Bristol pound: A tool for localisation, *Ecological Economics 146*: 273–281. https://doi.org/10.1016/j.ecolecon.2017.11.002.

Martin, R., Pike, A., Tyler, P. and Gardiner, B. (2015) Spatially re-balancing the UK economy: The need for a new policy model, *Regional Studies 50* (2): 342–357. https://doi.org/10.1080/00343404.2015.1118450.

Mason, P. (2015) *Postcapitalism: A Guide to Our Future*. London: Allen Lane.

Matthews, D. (2019) More than 50,000 people are set to get a basic income in a Brazilian city, *Vox*. [Online]. Available: www.vox.com/future-perfect/2019/10/30/20938236/basic-income-brazil-marica-suplicy-workers-party [Last accessed: 15 April 2020].

McFarland, K. (2017) Barcelona, Spain: Design of minimum income experiment finalized, *Basic Income Earth Network*. [Online]. Available: https://basicincome.org/news/2017/08/barcelona-spain-design-minimum-income-experiment-finalized/ [Last accessed: 3 July 2018].

McLaren, D. and Agyeman, J. (2015) *Sharing Cities: A Case for Truly Smart and Sustainable Cities*. Cambridge, MA: MIT Press.

Mitchell, S. and LaVecchia, O. (2018) *Amazon's Next Frontier: Your City's Purchasing*, Institute for Local Self-Reliance. [Online]. Available: https://ilsr.org/amazon-and-local-government-purchasing/ [Last accessed: 24 April 2020].

Moses, A., Smith, L., Fairbairn, C., Sandford, M. and Seely, A. (2016) *Future of High Streets, Debate Pack, Number CDP-0051*. London: House of Commons Library.

Mulvale, J. (2018) The cancellation of Ontario's basic income project is a tragedy, *The Conversation*. [Online]. Available: https://theconversation.com/the-cancellation-of-ontarios-basic-income-project-is-a-tragedy-101555 [Last accessed: 24 April 2020].

Nelson, A. (2018) *Small Is Necessary: Shared Living on a Shared Planet*. London: Pluto Press.

New York City (2015) *One New York: The Plan for a Strong and Just City*. [Online]. Available: www.nyc.gov/html/onenyc/downloads/pdf/publications/OneNYC.pdf [Last accessed: 11 May 2020].

New York City (2017) *Aligning New York City with the Paris Climate Agreement*. [Online]. Available: www1.nyc.gov/assets/sustainability/downloads/pdf/publications/1point5-AligningNYCwithParisAgrmt-02282018_web.pdf [Last accessed: 11 May 2020].

New York City (2018) *Global Vision: Urban Action: Voluntary Local Review: New York City's Implementation of the 2030 Agenda for Sustainable Development*. [Online]. Available: www1.nyc.gov/assets/international/downloads/pdf/NYC_VLR_2018_FINAL.pdf [Last accessed: 11 May 2020].

New York City (2019) *One NYC 2050: Building a Strong and Fair City, Volume 1 of 9*. [Online]. Available: https://onenyc.cityofnewyork.us/wp-content/uploads/2019/05/OneNYC-2050-Full-Report.pdf [Last accessed: 11 May 2020].

Ontario Government (2019) *Ontario Basic Income Pilot (Archived)*. [Online]. Available: www.ontario.ca/page/ontario-basic-income-pilot [Last accessed: 18 March 2020].

Painter, A. and Thoung, C. (2015) *Creative Citizen, Creative State: The Principled and Pragmatic Case for a Universal Basic Income*. London: Royal Society for the Encouragement of Arts, Manufactures and Commerce. [Online]. Available: www.thersa.org/globalassets/reports/rsa_basic_income_20151216.pdf [Last accessed: 11 May 2020].

Piketty, T. (2013) *Capital in the Twenty-First Century*. Cambridge: Belknap Press of Harvard University Press.

Porter, M.E. and Kramer, M.R. (2011) The big idea: Creating shared value, *Harvard Business Review*. [Online]. Available: https://hbr.org/2011/01/the-big-idea-creating-shared-value [Last accessed: 10 May 2020].

PricewaterhouseCoopers (2018) *Good Growth for Cities 2018: A Report on Urban Economic Well-Being from PWC and Demos.* [Online]. Available: www.pwc.co.uk/government-public-sector/good-growth/assets/pdf/good-growth-for-gities-2018.pdf [Last accessed: 24 April 2020].

Purcell, M. (2006) Urban democracy and the local trap, *Urban Studies 43* (11): 1921–1941. https://doi.org/10.1080/00420980600897826.

Reich, R. (2015) *Saving Capitalism: For the Many, Not the Few.* New York: Vintage.

Reusswig, F., Hirschl, B. and Lass, W. (2014) *Climate Neutral Berlin 2050: The Feasibility Study.* Berlin: Senate Department for Urban Development and the Environment, City of Berlin. [Online]. Available: www.berlin.de/sen/uvk/klimaschutz/publikationen/ [Last accessed: 18 March 2020].

Reynolds, J. (2017) Could Preston provide a new economic model for Britain's cities? *City Metric.* [Online]. Available: www.citymetric.com/politics/could-preston-provide-new-economic-model-britain-s-cities-3243 [Last accessed: 24 April 2020].

Rich, M.A. and Tsitsos, W. (2018) New urban regimes in Baltimore: Higher education anchor institutions and arts and culture-based neighbourhood revitalization, *Education and Urban Society 50* (6): 524–547. https://doi.org/10.1177/0013124517713607.

Roth, L. and Russell, B. (2018) Internationalism and the New Municipalism, *Resilience.* [Online]. Available: www.resilience.org/stories/2018-10-09/internationalism-and-the-new-municipalism/ [Last accessed: 17 March 2020].

Russell, B. (2019a) Beyond the local trap: New Municipalism and the rise of the fearless city, *Antipode 51* (3): 989–1010. https://doi.org/10.1111/anti.12520.

Russell, B. (2019b) Fearless cities municipalism: Experiments in autogestion, *Open Democracy.* [Online]. Available: www.opendemocracy.net/en/can-europe-make-it/fearless-cities-municipalism-experiments-in-autogestion/ [Last accessed: 17 March 2020].

Sennett, R. (2018) *Ethics for the City: Building and Dwelling.* New York: Allen Lane, an imprint of Penguin Books.

Sharp, D. (2016) Sharing cities: Why ownership, governance and the commons matter more than ever, *Shareable.* [Online]. Available: www.shareable.net/blog/sharing-cities-why-ownership-governance-and-the-commons-matter-more-than-ever [Last accessed: 18 April 2016].

Sheffield, H. (2017) The Preston model: UK takes lessons in recovery from rust-belt Cleveland, *The Guardian.* [Online]. Available: www.theguardian.com/cities/2017/apr/11/preston-cleveland-model-lessons-recovery-rust-belt [Last accessed: 17 March 2010].

Singer, C. (2016) *The Preston Model, The Next System Project.* [Online]. Available: https://thenextsystem.org/the-preston-model [Last accessed: 24 April 2020].

Slaper, T.F. and Hall, T.J. (2011) The triple bottom line: What is it and how does it work? *Indiana Business Review 86* (1). [Online]. Available: www.ibrc.indiana.edu/ibr/2011/spring/article2.html [Last accessed: 24 April 2020].

Smith, D. (2008) The politics of studentification and '(un)balanced' urban populations: Lessons for gentrification and sustainable communities? *Urban Studies 45*: 2541–2564. https://doi.org/10.1177/0042098008097108.

Standing, G. (2017) *Basic Income: And How We Can Make It Happen.* London: Pelikan Books.

Streeck, W. (2016) *How Will Capitalism End? Essays on a Failing System.* London: Verso Books.

Thompson, M. (2020) What's so new about New Municipalism? *Progress in Human Geography*: 1–16. https://doi.org/10.1177/0309132520909480.

9
TRANSITIONING TO ETHICAL CITIES

The Great Khan's atlas contains also the maps of the promised lands visited in thought but not yet discovered or founded: New Atlantis, Utopia, the City of the Sun, Oceana, Tamoé, New Harmony, New Lanark, Icaria.

Kublai asked Marco: You, who go about exploring and who see the signs, can tell me toward which of these futures the favoring winds are driving us.

Italo Calvino, Invisible Cities, p. 147

Ethical realism

A perennial paradox of the urban condition is that, on the one hand, it is compelling to take a deeply pessimistic view of the prospects for ethical city futures when forces of conservative reaction and capital fill the viewfinder. As an antidote, a quote from the late Mark Fisher reassuringly suggests how:

> the very oppressive pervasiveness of capitalist realism means that even glimpses of alternative political and economic possibilities can have a disproportionately great effect. The tiniest event can tear a hole in the grey curtain of reaction which has marked the horizons of possibility under capitalist realism. From a situation in which nothing can happen, suddenly anything is possible.
>
> *(Fisher 2009, pp. 80–81)*

Events such as the GFC and the Covid-19 pandemic illustrate how fragile capitalist economies and ways of urban life are. A positive futurist perspective seems even more plausible if one suspends disbelief and surveys the plethora of innovations around new municipalism, inclusive economies and decarbonisation (see Chapter 8). The GFC and pandemic each represent in contrasting ways both challenges to the brutal normalcy of capital and opportunities to move in constructive directions away from the status quo. Perhaps ethical realism could replace capitalist realism. This infers that there are objective, knowable ethical truths that align with their scientific counterparts (Werner 1983; FitzPatrick 2015; Metcalf 2015). Ultimately, it is reality reconfigured in the context of ethically symbolic activities calculated to do away with the invisible barriers constraining our dreams, thoughts and actions under capitalist realism (Fisher 2009, p. 16; Duncombe 2019).

Amidst the sustainability, economic, financial and representative governance crises that constitute this defining moment in the twenty-first century, it is perhaps inevitable that transitioning has become widely applied across social and political sciences – and beyond. For many, this involves a search for promising ways out of the individualistic, winner-takes-all charade that constitutes late capitalism (Giridharadas 2019). A critical assessment of transitions optimism is necessary, and this is not inconsistent with advocacy for relentless attempts to maximise and practice awareness, learning and action-based solidarity and collective agency. On the premise that change starts from contemporary constraints, any transition towards ethical cities immediately confronts multiple, reinforcing incumbencies. Most recently, several decades of policy settings to favour capital accumulation and widening inequalities marked the ultimate demise of the reformist Third Way project to accommodate capitalism. As highlighted in Table 9.1, the Third Way as a political philosophy was meant to occupy the radical centre of politics, to promote cosmopolitanism and the new democratic state (Giddens 1998; Giddens 2000). In recognition of the failure of the project to address deepening working-class poverty and environmental destruction, a more overt stance towards the excesses of neoliberalism has emerged in the centre-left across the Global North, marked by priorities to decarbonise economies while addressing inequality and persistent poverty. From these have re-emerged ideas of re-nationalisation of key social assets (health, infrastructure, education), universal basic income and community wealth building to provide a social safety net during the transition for individuals, communities and cities. This has not (yet) attracted sufficient disgruntled working-class voters across the UK, USA and other western democracies, but it has confronted the far-right onslaught of racism and anti-multicultural populism.

Meanwhile, the working class bears the brunt of impacts associated with the digital society heralded by the gig economy, bringing new assaults on ordinary families, rolling back hard-fought basic rights to employment, working conditions, healthcare and education. All indications are that a collapse in job security and social fabric may well spiral much further, if left unattended by governments at all levels. Artificial Intelligence and casualisation are powerful allies and very profitable in the short term. Meanwhile, post-truth media has opened up new opportunities for antisocial action, and the lack of impartiality of press and the judiciary has eroded the collective faith in justice and integrity. In aligning with emancipatory politics tied to the reinvention of democracy and integrated approaches to climate change and inequality, ethical realism requires a distinct philosophical framework (some preliminary ideas are shared in Table 9.1). Again, it should be highlighted that as a processual rather than didactical concept, movement towards the ethical city is measured by the open, transparent and accountable governance and progress in tackling inequality as well as environmental restoration.

What are the prospects for such emancipatory change and what alternative futures might be on their way? At the turn of the century the ecological economist Robert Constanza speculated on four possible futures (Costanza 2000). He presented a 'Star Trek' scenario as an optimistic outcome for pro-technology advocates, with a more Armageddon-like pessimistic version termed 'Mad Max' after the movie of that name. The remaining two scenarios were technologically sceptic and termed 'Big Government', where capitalist organisations are found wanting and reigned in and 'Ecotopia', where collective agreement is reached on decarbonisation and the conservation of ecosystems. Similarly, the Australian perma-culturalist David Holmgren shared four future scenarios around techno-explosion, techno-stability, managed descent and collapse. On the positive side, he envisioned a world where technology either facilitates continued exponential growth or conversely the attainment of a steady-state economy. In his worst-case scenarios, the world would face systemic collapse. The most creative and realistic scenario for Holmgren, however, implied a gradual energy descent very similar to most decarbonisation scenarios produced by the IPCC (Holmgren 2009). Focusing on the question of what life will be like after capitalism, Peter Frase also posits four scenarios: two heavens in the form of communism (typified by egalitarianism and abundance) and rentism (hierarchy and abundance) and two hells in the form of socialism (egalitarianism and scarcity) and extremism (hierarchy and scarcity) (Frase 2016). Rentism is a simple extrapolation of contemporary trends where abundance exists but is monopolised by a small elite. Communism for Frase is a society that is productive and egalitarian (not authoritarian) with automation technology providing the potential for a post-work, post-scarcity and post-carbon world.

If we extrapolate these scenarios to cities, we would inevitably find a few with the potential to be become Star Trek with high technology, seamless mobility and high levels of public safety (think of Tokyo here). Others might increasingly become fortresses, with gated communities and spiralling scarcity, inequality and crime – the next level of new military urbanism (Graham 2010). Another group might become havens, sanctuaries and lifeboats for migrants and other victims – the emerging fearless cities movement (Russell 2019). A fourth group might by various definitions face collapse and become failed, feral and fragile cities (de Boer 2015; Muggah 2016). The latter is very much in line with projections made by James Howard Kunstler in *The Long Emergency* published back in 2005. He was concerned that we were overdue a pandemic that would paralyse socio-economic systems, interrupt global trade and bring down governments (Kunstler 2005). More recently, Ravetz (2020), drawing on the work of Inayatullah and

TABLE 9.1 Characteristics of ethical realism compared to the Third Way, Old Left and New Right perspectives

Classical Social Democracy	Neoliberalism	Third Way	Ethical Realism
The old left	The new right	The radical centre	Emancipatory politics
State involvement in social and economic life	Minimal government	The new democratic state (the state without enemies)	Engaged experimentation with the demos and respect for general will of the people
The State dominates civil society	Autonomous civil society	Active civil society	Networks of actors
Collectivism	Market fundamentalism	The democratic family	Collective agency
Keynesian demand management, plus corporatism	Moral authoritarianism, plus strong economic individualism	The new mixed economy	Low-carbon, inclusive, democratic economy
Confined role for markets: the mixed or social economy	Labour market clears like any other	Equality as inclusion	Community wealth, prosperity without growth
Full employment	Acceptance of inequality	Positive welfare	Addressing inequality and persistent poverty
Strong egalitarianism	Traditional nationalism	The social investment state	Universal basic income
Comprehensive welfare state, protecting citizens from cradle to grave	Welfare state as a safety net	Philosophic conservatism	Disruptive modernisation
Linear modernisation	Linear modernisation	Ecological modernisation	High ecological consciousness
Low ecological consciousness	Low ecological consciousness	The cosmopolitan state	Feminisation of the state
Belongs to a bipolar world	Realist theory of international order	Cosmopolitan democracy	Caring democracy
	Belongs to bipolar world		Belongs to a multipolar world

Source: First three columns derived from Giddens (1998).

Black (2020), posited four post-pandemic scenarios; a rapid return to business as usual; dramatic societal transition moving beyond old economic values and governance systems; and two more dystopic scenarios where pandemic conditions extend through repeated resurgences over several years (Kissler et al. 2020). In one, rapid societal transition creates new communities of hyper-networked isolationists and in the other, surveillance/disaster capitalism and power grabs result in some cities becoming zones of exclusion and oppression (Ravetz 2020). While such scenarios are useful heuristics for imagining futures, our main focus is the mechanisms by which urban change occurs and how some sorts of transformation might lead to more desirable ethical cities; emphasising that mechanisms to reinforce positive transformation require an ethical orientation be articulated, empowered and maintained, rather than a teleological end-state. Recent and current ideas on just how deep urban changes might occur and sustain are the main topic of this chapter. First, a short summary of some of the key points throughout the preceding chapters provides a baseline for discussion of this ultimate challenge.

Revisiting the key points

In Chapter 1, the concept of the ethical city was presented as an antidote to popularism, individualism, tribalism, neoliberalism, nationalism, local parochialism and the fatalism associated with acceptance of humans as cruel or late capitalism as predestined. Rather, the ethical city calls for active positioning towards equity, transparency, accountability, environmental sustainability, restorative and climate justice. What is more the ethical city is not an endpoint/blueprint; it is an alignment, trajectory or proclivity. It exhibits respect for human and non-human rights, social inclusion, dignity and community engagement. These are built upon a foundation of transparent, democratic, accountable and ethical governance. The ethical city is actively opposed to inequality as an antisocial phenomenon. It is distinct from the notion of the ethical individual (Bloom 2017) and accepts that we are collectively shaped by the codes, rules and socially sourced truths that surround us, and that, in turn, we shape them. Hence, the ethical city is necessarily an endeavour through which collective agency acts as a conduit, not replacement, for individual action. There is a clear role for advocacy and recognition of the importance of ethical practice.

The role of universal rights is the focus of Chapter 2. As well as bringing rights into alignment with the ethical city, we posit the ethical city in contrast to several decades of scholarship on the right to the city and chart reinforcing and diverging threads between these ideas. In Chapters 3 and 4, ideas of ethical codes and approaches and ethical practice of individuals involved in urban decision-making are introduced, respectively.

At the urban scale, questions were posed in Chapter 5 about how we get to know the ethical city via the lens of assessment. Through the ubiquitous mantra of plan, do, act, check, a plethora of iterative planning tools have arisen, designed to assist urban communities in setting out plans, reporting progress, altering course and determining resource allocation accordingly. The SDGs are a clear frame of reference for such evaluation and Voluntary Local Reviews of various sorts provide an example of this contemporary assessment tool (although still very much work in progress). In Chapter 6, an account is provided of the differences we might expect to see between cities that take an active orientation towards inclusion and sustainability and those that take a laissez-faire approach, defaulting to a neoliberal, competitive, unsustainable city mindset. The dynamics of disruption in cities is brought to the fore in Chapter 7, as cities, at once both people and place, and at once both obdurate yet also in flux, remain seemingly resistant to all attempts to describe or control, let alone to predict their destiny. In Chapter 8, we turn to an assessment of the myriad of practical on-going innovations intended to change cities in more sustainable directions, with a particular emphasis on grassroot, locally oriented, intentional interventions.

This leads us to the present question: How might the sort of deep changes articulated towards the end of Chapter 8 take hold and become more widespread? Moreover, how can interventions reframe relations, social, material or otherwise, in meaningful ways to reconfigure cities along more ethical lines as we have described throughout this book?

Interruptions, disruptions and transitions

One thing is sure, interruptions and disruptions (whether planned or accidental) to current modes and sites of capital accumulation will continue to alter the urban fabric. This includes rapid technological progress, extreme climate change events, natural disasters and pandemics. There are also associated societal responses in the form of intentional strikes, other forms of interruptions, civil disobedience and protest that impact on the means of production. All of the above offer the opportunity to adapt, rethink and reorder, for better or for worse, for good or evil, temporarily or on a more extended basis. Such moments of insecurity and instability provide points in time where incumbency is potentially weakened or challenged as found recently in examples around energy and climate change (Ko et al. 2019; Solecki et al. 2019). However, the test of relative robustness of a society is whether or not these transformative events can be nudged in a positive direction, rather than becoming victim of disaster capitalism resulting in the introduction of questionable policies and practice – for instance highly authoritarian modes of government (Klein 2007). Interruptions to consumption, whether due to economic collapse or voluntary/organised mass boycotts of goods and services may also prompt challenges and adjustments, more or less significant in their outcomes. Such interventions to production and consumption may of course occur in concert with each other. Moreover, there is a third form of intervention that we simply describe as collectively reorganised forms of production and consumption and parallel economies of non-profit, shared, or otherwise, based upon mutual aid. There are myriad on-going experiments dedicated to testing new modes of urban living outside the market and within natural carrying capacities. The question is whether and how these might replace or somehow ameliorate dominant modes of urban life in late capitalism that are unsustainably consumptive and alienating.

Socio-technical transitions scholarship originates from a central assumption that science, technology and society are co-produced and in a constant state of tension, reinforcement and change (Bijker et al. 1987). Evolutionary economics and the sociology of innovation together inform ideas of current regimes, being bundles of institutions, rules (formal and informal), capabilities and conventions that hold together ways of doing things, and therefore, patterns of inequality and unsustainability. In turn, this provides the basis for understanding how people and institutions shape actions. It follows from this conceptualisation that socio-technical transitions require a combination of novel technologies and social changes, whether these are concerted, planned and designed through, for example, deliberate rulemaking or prompted by other societal factors. Perhaps the most well-known description of this process is the Multi-Level Perspective (MLP), which envisages transitions as successful disruptions to an incumbent regime caused by the development of a new niche, supported by external factors and accompanied by an enabling landscape (Geels 2010). A good example to illustrate this is the way that renewable energy is gradually moving from a niche technology to become a dominant form of energy production, but still with a very long way to go. Thus, transitions are viewed in some quarters very much as complex, contested, non-linear processes occurring at multiple levels with niche and emergent innovations shaping new socio-technical regimes, all the time having to overcome in-built systemic resistance (Geels 2005; Geels and Schot 2007; van den Bergh et al. 2011; Geels 2014). A clear benefit is in their application to bridging analytical approaches in decarbonising society (Geels et al. 2016).

The spatial and urban turns in transitions have produced a burgeoning literature specifically focussed on how such socio-technical transitions processes might proceed within cities and across urban terrain (Moore et al. 2018). Urban transition labs or urban living labs have emerged based on the observation of emission reduction practices in several European cities (Nevens et al. 2013; Evans and Karvonen 2014; Voytenko et al. 2016). These experiments can be thought of as interventions that support learning around the delivery of sustainability goals for cities, including providing research opportunities to reveal how power and agency are orchestrated in order to get desired results (Bulkeley et al. 2016; Caprotti and Cowley 2016). They can be implemented across a range of thematic areas (mobility, energy, information and communication technologies etc.) and led by either government, private entities, academic institutions and civil society/not for profit actors. This raises the question of the role that urban communities, and, for that matter, city governments and urban elites play in shaping or directing socio-technical transitions, rather than being passive recipients of transition initiatives (Hodson and Marvin 2010). City actors can support niche innovations that gradually emerge to transform urban socio-technical regimes (Geels and Schot 2007; Newton and Bau 2008). For instance, using the example of renewable energy again, instead of passively consuming brown energy from a regional grid, cities seeking to decarbonise can harness new technologies to generate as much renewable electricity as possible from within their territorial boundaries. This leads to ideas like the solar city whereby large-scale photovoltaic infrastructure is installed in the city, with appropriate policy and financing mechanisms. The feasibility of this approach has been examined with reference to cities as diverse as London, Munich, New York, Seoul and Tokyo (Taminiau et al. 2015; Byrne et al. 2016). Indeed, over 100 cities are reported as predominantly powered entirely by renewable energy sources (Hunt 2018). Many cities, nevertheless, are struggling to unpack the complexities around what low carbon actually means, who is involved in the transition, how it should unfold and how to recognise the most effective pathways towards the attainment of this transition. Clearly, emphasis should not be solely on technical and infrastructural shifts, but also requires dramatic reconfigurations in how we govern patterns of development with transformative political and economic implications (Luque-Ayala et al. 2018). This involves not only judgements of fairness, justice and inclusiveness, but also, inevitably, questions of power.

Cognizant and responsive ethical city

From an ethical cities perspective, a renewable energy powered city is not in itself a sufficient result. In many jurisdictions, policy settings supporting domestic-scale photovoltaic renewables transitions have come not only from the city, but also from regional, state or national governments and have tended to be funded via a levy on retail electricity prices. As discussed in Chapter 1, low-income renters who cannot afford matching funds to claim these rebates, and whose landlords in any event do not want the systems to be fitted, are paying higher bills as they effectively subsidise well-off homeowners of large houses with large rooftops that can accommodate sizeable domestic PV systems. For instance, research on the solar tax credit in Hawaii found that homeowners with a medium income and above benefit most from PV systems (Coffman et al. 2016). Thus, richer households benefit financially and feel good about their emission reduction efforts, while disadvantaged households pay the price. In an ethical city, the introduction of renewable energy must go hand in hand with accountability and inclusive attempts to end inequality. On this basis, a technological approach to decarbonisation which does not account for ethical concerns around who benefits and who loses is a regressive step. As this case illustrates, the making of interventions and interruptions that promote just urban transitions is the key. A system of power relations that supports regressive policymaking is as much a barrier to ethical cities as the collective will to decarbonise and understanding of what this constitutes. On the matter of collective will and agency, in times of populism and anti-environmentalism, this cannot be taken for granted. During times of disruption, change is particularly inevitable and coalitions to shape the future in more inclusive and sustainable directions are pitted against darker, self-serving forces seeking to take advantage of anxiety, hopelessness, poverty and misery in calculated ways that will fuel more hatred, division, and climate change (Klein 2007).

Much is made of the idea of co-design to promote democratic policy and action towards sustainable urban transitions. While this is laudable, there are clear conditions when power is to be accounted for. While co-design and a collective approach hold out the possibility of ameliorating the worst effects of dominant interests, power dynamics are more sophisticated than simple representation might suggest. As Avelino (2009) has shown, such ideas can risk naivety, as revealed in a study of the micropolitics of transitioning in the Dutch transport sector. The ethical city, therefore, must also be a cognizant and responsive city. A seat at the table for each stakeholder is not enough; in democratic decision-making, elites still invariably dominate the contest of language and ideas. With this in mind, Avelino et al. (2016), Meadowcroft (2011) and Avelino and Wittmayer (2015) all argue for a deeper consideration of politics in transitions processes. In this vein, the public fightback against the corporate agenda surrounding smart cities could be portrayed as a classic David and Goliath situation, where highly paid corporate lawyers are pitched against spare time residents in a battle over fine print contracts that will ultimately determine whose data gets sold to who, at what price and who will get to know (or not). In response, as mentioned in Chapter 7, the city of Barcelona coined the term 'technological sovereignty' in October 2016 as an attempt to reset the power dynamic in such transactions (Kitchin 2019). A set of ethical digital standards have been developed by the city promoting the use of free software, interoperability of services and systems and the use of free standards. The approach involves designing public services as digital by default. But more importantly, this should involve placing people at the centre of the design process. Online services must be built in the most agile and open manner possible; they must be simple, modular and interoperable to avoid dependencies on vendors and providers of specific proprietary solutions. This should be tied into transparency around information and data sovereignty so that residents and government know what data is collected, how it is used and who owns it (Barcelona City Council 2016).

Actioning transitions to the ethical city

Transitions are contingent projects with unknowable outcomes. The management of transitions (Loorbach 2007), like ideas of ecological modernisation before them (Mol and Spaargaren 2000, Barrett 2005) have been problematised on a number of grounds, not least that they tend 'to give primacy to the levers of government or, at least to large institutions, in instigating and convening transitions' (Horne 2018, p. 7). Since such institutions are replete with the advantages of power outlined above, the dangers of path dependencies abound in transition management projects. At the very least, urban change can be described as unmanageable and unpredictable. Nevertheless, it is possible to speculate on the sorts of actions that may be more promising in promoting change, with these severe contingencies in mind. We propose six interlinked opportunities for promoting urban transitions to ethical cities, namely:

- Mediate the means of production;
- Mediate consumption;
- Reconfigure modes of exchange;
- Processes rather than endpoints;
- Overcome fear and embrace the inevitable; and
- Build alliances, share knowledge.

Mediating the means of production

As illustrated in Chapter 7, there is no shortage of analysis demonstrating the negative consequences of unfettered market-based automation on the future of work. In addition, AI presents a new opportunity to reorder the city. Under the 'BAU' scenario of entrenched deregulated capitalism, highly skilled and advantaged social groups will thrive while those lower in relevant skills, wealth and status will be cast aside. Put simply, if labour is needed to do this then there will be work; if not, there won't. Structural funds, support or investment in retraining, alternative industries, or other conventional means to provide a just transition through neo-Keynesian measures are selectively considered according to the power structures and dynamics at play in any given case. Voices and actions of dissent are proven agents in such struggles, with attempts to ensure new technologies are adopted for the benefit of a broader urban public rather than the elite. For example, the City of San Francisco sought to ban the use of facial recognition technology arguing that it is incompatible with healthy democracy (Conger et al. 2019). This contrasts sharply with the situation in London where the police began using facial recognition technologies in January 2020 ignoring concerns expressed by civil liberty groups (Vincent 2020). An independent review of test deployments of this technology by researchers from the Human Rights Centre at the University of Essex concluded that the deployments would be considered unlawful if challenged in court. Further, the researchers called for meaningful engagement and debate prior to the introduction of such technologies (Fussey and Murray 2019). Elsewhere, with respect to another emergent technology, the city of Pittsburgh called for an ethical road map in advance of widespread autonomous vehicles adoption that ensures companies involved act in a socially responsible manner, including sharing their data and giving back to the affected communities (Thomas 2017). Likewise, there are calls for cities to lead in directing the adoption of AI and automation – just as they have in fighting climate change and protecting the rights of migrants (Miller and Asare 2018). The main concern is that some cities will be more significantly impacted than others as demonstrated by a 2017 study by the Institute for Spatial Economic Analysis. The researchers found that certain low-wage metropolitan areas in the US risk losing up 63 per-cent of their jobs to automation by as early as 2025 (Chen 2017). Similar research undertaken in the UK found that one in four jobs in cities in the north of the country are at risk from automation, potentially deepening political divides in an already post-Brexit stressed nation. In this study, the total number of jobs lost to automation and other technological disruptions in British cities by 2030 was estimated at 3.6 million (Centre for Cities 2018, p. 13).

How might interruptions that reconfigure the means of production be directed towards more just outcomes? A narrow, tactical response is to seek to compete with other cities for the higher value jobs, by investing in education, rolling out retraining and upskilling programmes and seeking to attract talent through marketing and positioning. This may or may not result in more employ-ment opportunities locally, but in any event will promote uneven outcomes – winners and losers – at various scales, thus exacerbating current inequalities and related forces of reaction. Rather, cities could take a more collective approach and coordinate the development of innovation hubs to support local social enter-prises and start-ups (Miller and Asare 2018). In the end, this may come down to a very simple question elaborated by the London-based Centre for Cities (Magrini and Clayton 2018): Can cities outsmart robots and AI or, more pointedly, can they outwit the people and corporations behind the automation and AI agenda? To achieve this, cities are going to need to engage with a growing set of stake-holders to navigate the ethical dimensions of a whole range of issues including humans vs. machines, public vs. private, virtual vs. real, and consumer vs. citizen vs. user (Pitroda and Miailhe 2017). Many ethical concerns have already been flagged in relation to AI around data ownership, distributed sensing, privacy and surveillance. There are no easy solutions and it is going to take a concerted effort by both urbanists and urbanites to respond to these concerns (Cranshaw 2013). However, a starting point is a four-fold approach that: (1) links to long-standing campaigns and models of labour organisation; (2) forms direct democratic organ-isations of workers, (3) seeks justice and upholds basic rights and (4) uses legal, solidarity and other powers to exercise agency. Nowhere is this clearer than in relation to deep decarbonisation of the economy which should be devised so as to avoid impacting disproportionately on low-income households. Germany is leading the way in this transition with plans to phase-out coal production by 2038, allocating $45 billion for economic restructuring, job creation and work-force retraining (Elliott 2019). Other countries, including Canada, Spain and South Africa are also involved in stakeholder dialogues on how best to manage a just transition away from fossil fuels.

Mediating consumption

Inevitably, conventional measures of economic growth present long-held and obdurate barriers to cities seeking to curb consumption (rather than switch to renewables or invest in retrofitting to promote energy efficiency, both of which offer prospects for new jobs at least in the short term). How can cities cut consumption while maintaining their role in global market-based relations? Here lies an obvious role for communities acting collectively to address the environmental impacts of consumption by reducing the environmental burden associated with going about their daily lives. While social structures are an important constraint upon consumer action, broader socially engaged interventions around climate mitigation, for example, rather than attempts at individual behaviour change can provide an impetus for inclusive and meaningful transition. Over a single generation, diets, housing, consumption and travel, energy sources etc. will all need to transform beyond recognition; this will not occur through voluntary lifestyle 'choices', while business continues to sell us meat, jet-set aspirations and countless other sources of social challenge (Institute for Global Environmental Strategies 2019). As many (including the authors) have asserted, social change is the only way that society will be transformed, including coordinated efforts of reconfigured governance. As Kolbert (2015) pointed out about Zurich's long-standing efforts to encourage residents to adopt and live within a '2000 watt' limit of energy consumption (City of Zurich 2011):

> Here's a piece of bad news I will share: since adopting this target, Zurich has made almost no progress toward it. Setting abstract goals isn't difficult. The US does this all the time, and so does Canada. The hard part is fulfilling them.

A 2007 survey of over 3,300 Zurich residents revealed the scale of the challenge with only 2 percent exhibiting a rate of energy consumption below 2000 watt, leading the researchers to conclude that systemic changes were essential in order to reach the target by 2050, including supplying 80% of total energy from renewable sources (Notter et al. 2013). Subsequently, the City of Zurich, in common with many other cities, has apparently achieved primary energy reductions and CO_2 emissions reductions (Figure 9.1), through coordinated structural changes to infrastructure, energy supply, buildings and mobility (City of Zurich 2016). However, there are many rapidly growing cities around the globe with high per capita emissions including Shanghai, Beijing and Tianjin in China (Baeumler et al. 2012).

Reconfiguring modes of exchange

Reconfiguring social economic arrangements includes shifts from capitalist consumption towards social or non-market forms of exchange, and prosumption (production by consumers), where chains of production and consumption are shortened or brought together. The switch from commerce-based activity to mutual aid and non-financial exchange may also include some forms of community self-provisioning. Whilst various online applications promote ride sharing

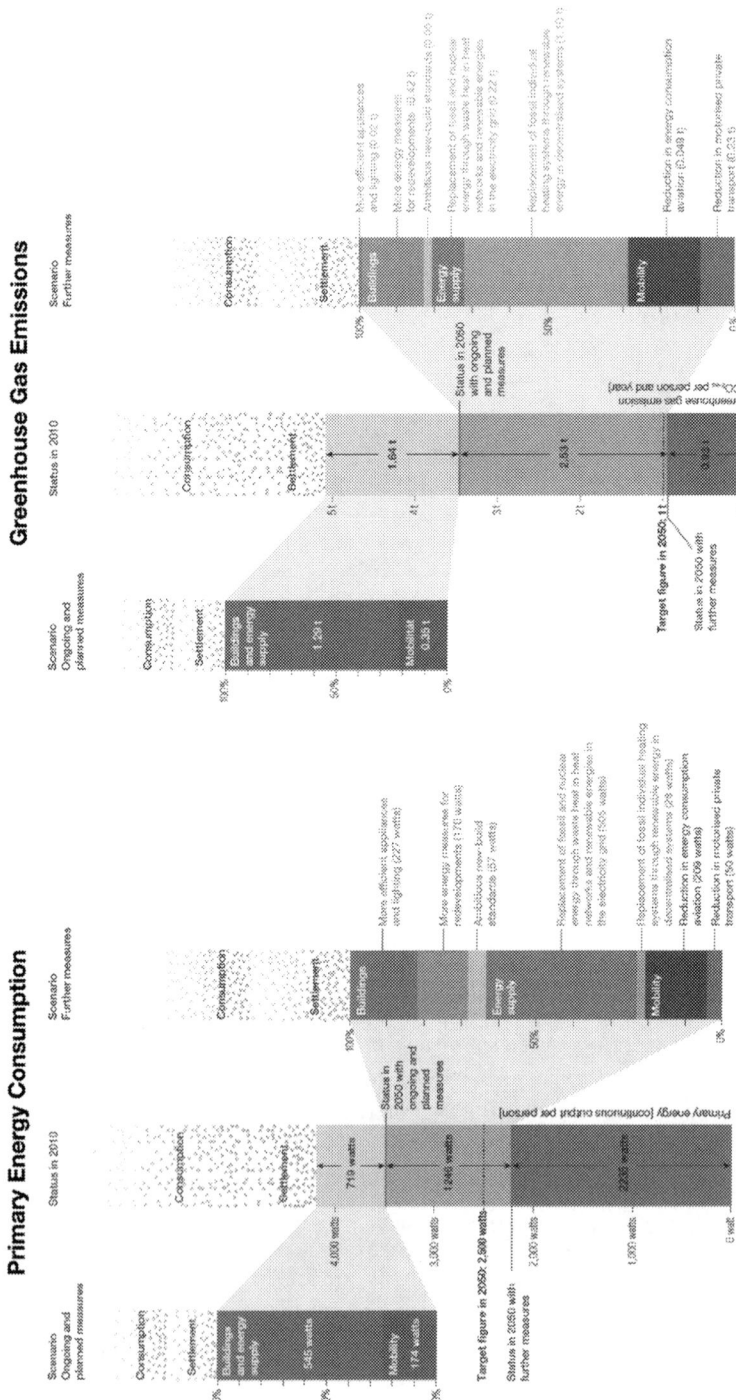

FIGURE 9.1 City of Zurich potential measures to meet 2050 targets

Source: City of Zurich 2016.

or house sharing, for example, these have infamously been captured by capitalist models of accumulation, bringing new waves of casualisation and gentrification in the process. This has resulted in push-back from many cities. Paris, Barcelona and Santa Monica have all implemented strict policies to regulate Airbnb. Further, Uber was at one point banned in London due concerns around passenger safety (Tun 2020; Thomson et al. 2019). The bigger challenge, however, is for cities to discover ways to harness such technologies for more sustainable and ethical ends. Collectivist approaches such as the shared economy and collaborative consumption are promising ways to weather the contemporary techno-economic storm and economic down-turns, but only with strict provisos associated with new business models of social enterprise that seek to reframe commodity chains as social assets. Initiatives with promise are those that counter the shift towards monetisation and marketisation of everything and instead allow for coexistence of another economy built on open reciprocity, good conduct and caring for fellow residents, as well as different notions of wealth and prosperity (Jackson 2009).

Strategies reconfiguring existing modes of exchange that aim to coexist with capitalism, albeit in protected spaces, often build upon what already exists and take advantage of new technology as it emerges. For example, the local currency in Bristol, UK (the Bristol Pound) has utility in aiming to keep economic activity local, but it has encountered a shift more broadly towards digital monetary systems. While of course it has the UK Pound to co-exist with, it can also be a digital currency, with added benefits of providing quasi-economic exchange as tokens or credits. These could be either granted by the city government to promote advancements in Bristol's efforts towards the SDGs (e.g. for cycling, avoiding rush-hour commuting, taking your own cup to the café or checking on your elderly neighbour; or to support a skill-swap/exchange system – hairdressing for baby-sitting, gardening tools for help with fence painting etc.). The advantage over Airtasker type apps is that this avoids formal monetary exchange and so is less exposed to profiteering through the formal economy. The idea of such novel, local, not-for-profit digital currencies, as researchers in Spain found, is that they promote a sense of belonging and personal empowerment and encourage solidarity and altruism (Gimenez and Tamajon 2019). Such outcomes are very much in line with the emphasis on intrinsic values that we argue is central to the ethical city (Crompton and Weinstein 2015).

Processes rather than endpoints

As introduced in Chapter 1, the ethical city is not a blueprint, but a process. The didactic, immutable rules that plans and blueprints infer (Bregman 2017) are anathema to the co-produced ever-a-work-in-progress ideals of the ethical city. Bregman invokes the example of the City of the Sun, a work of Italian poet Tommaso Campanella. In this city, every aspect of private life is controlled by the state. Individual ownership is prohibited and everyone is obliged to love everybody else. 'What's more', Bregman explains, 'every person is monitored by a vast network of informants. If someone commits a transgression, the sinner is verbally browbeaten until they are convinced of their own wickedness' (ibid., p. 12). While the blueprint can clearly turn authoritarian, this does not mean we are left only with the DIY city where anything goes. A clear set of rules about governance, accountability and representation is essential; as is an unwavering focus upon tackling inequality and improving environmental quality. An ethical city orientation provides space for new thinking, politics and action. As discussed in Chapter 1, ethical democracy is about rebuilding the demos as a means to oppose corruption and injustice. Without mass awareness and engagement there will be no overcoming of existing power dynamics associated with urban change processes. Taken together, these ideas add up to a call for practice that is ethical and processes that are just. The ethical city is not an entity, but a string of connected actions that together add up to new routines, patterns and methods aimed at bringing about positive change. Just as urban systems are in a constant and dynamic state of tension and change, so the ethical city reflects these realities and is in continuous need of ethical revision.

Overcoming fear of the inevitable

For cities, whether these are people or bureaucracies, their subnational frame makes them constantly subordinate to national and global regulation, deregulation and market forces. A heavy cloak of permanence and inevitability hangs over the capitalist project. Yet, the starting point in contemplating the future is the certainty that capitalism will end (Frase 2016). Change is inevitable and can be good or bad. The end of capitalism is inevitable since it will evolve in different directions. In this regard, fear of the end of capitalism is primarily a construct of those who have most to lose – namely, the 1 percent.

Overcoming fear may occur in various ways. Pandemics or other major shifts associated with extreme weather and climate change might have impacts that are so violent they to push society past a tipping point where existing capitalist structures break down irreparably and change is sustained. We have already shared alternative scenarios on how this could herald Armageddon or progressive change. Ethical city actions prepare the social fabric and practice for progressive alternatives, which at the same time help overcome fear of the unknown. A means to reduce fear is to collectivise and network amongst like-minded communities. The contention here is that neighbourhoods, towns and cities around the world can work together, in the face of drastic change, to defend human rights, democracy and the common good (Gellatly and Rivero 2018). At the 2018 Fearless Cities Summit in Brussels reference was made to two types of fear: one that paralyses and prevents movement and the other that makes you jump or run (Huberlant and VerKamp 2018, p. 24). The only way to handle the latter is through movement. A set of feminist intrinsic values were proposed to guide this movement including collaboration, dialogue, horizontality, learning-by-doing, trial and error, process-before-outcomes, allowing doubts and focusing on the concrete and practical. The aim should be to create a safe mental and social space within which individuals become fearless and can take a leap forward (ibid., p. 7).

A range of fears abound around automation, AI, and the next phase of economic upheaval and its implications for decent work, employment and inequality. The problem is not new technology; it is about how it is abused. In embracing fully automation, we can think about how it can help reduce mundane work while improving working lives, providing opportunities for self-realisation, providing universal basic income and taking advantage of technological and other disruptions in order to achieve a progressive alternative agenda. Its stance is outlined by the political theorists Nick Srnicek and Alex Williams in their 2015 book *Inventing the Future: Postcapitalism and a World Without Work*. The authors suggest that the way forward is to think big and 'rather than settling for marginal improvements in battery life and computer power, progressives should mobilise dreams of decarbonizing the economy, space travel, robot economies – all the traditional touchstones of science fiction – in order to prepare for a day beyond capitalism' (Srnicek and Williams 2015, p. 430).

Build alliances, share know-how

Fortunately, collectivising resistance and coordinating transition efforts between and across cities is a growing trend. The Urban Transitions Alliance, a network of cities based in the USA, Germany and China, has examined the potential for transitions across four core sectors: infrastructure, energy, mobility and social. These cities argue that while past transitions occurred without choice (e.g. hollowing out of former manufacturing industrial cities), cities today are embracing transitions that lead to new opportunities but in a way that prioritises the needs of their local populations (ICLEI 2019). The idea that individual, cash-strapped cities are powerless to stand up to mega-tech companies who control global infrastructures goes some way to explaining the growing plethora of city networks, where cities can share knowledge between each other as a means to counter the reach of global corporate power. Two cautions are relevant, however, in attempts to build alliances and share knowledge on urban transitions towards ethical cities.

The first caution relates to the need for coherence and integrity of purpose in any alliance. As we have previously mentioned, single purpose projects such as low carbon cities can fall into the trap of exacerbating inequalities or undermining inclusion if the focus is solely upon progressing decarbonisation, and this takes place within the mode of uncritical ecological modernisation. Networks that are productive must therefore be linked by common ethics. Aware cities, cooperating on specific, integrated projects in line with ethical principles and incorporating ethical processes would provide an antidote to the notion of the competitive city, where all cities are encouraged to compete for the same resources – even to the death – as in the United States, where cities where forced to bid against each other for ventilators and personal protection equipment during the early stages of the Covid-19 pandemic.

The second caution surrounds the flawed idea that 'community' is necessarily a virtue, and that public engagement is declining in the digital era, and that, therefore, restoring social contract is a good thing per se. As introduced in Chapter 1, in advocating for community, we also need to be aware that creating or reinventing community by being exclusionary and parochial is something that fundamentalists can be very good at, and this takes society in the opposite direction to the ethical city. Alliances are only socially productive and ecologically sustainable to the extent that they enshrine and embody ethical city principles of universal rights and political consciousness. This brings us back to Kant's categorical imperative, as discussed in Chapter 3, which in this context would imply that communities within the ethical city and within networks of ethical cities should be universally open and inclusive.

Conclusion

The viability of the ethical city in the twenty-first century will depend upon contingent processes personified by actions of solidarity and vigilance to address inequality and poverty; governance/representation to bring about an engaged demos and sustainability to overcome the climate and ecological crisis. It infers strengthening global networks of practice and countering ideological barriers and fake news with real, collection action, self-provisioning education, building networks of knowledge and commonality. Working outside the market where possible, but avoiding cliques and communities of difference, ethical city projects span altering production (rights and action), mediating consumption and reframing exchange (non-monetary mutual aid etc). Those who seek to make ethical cities are building bridges of hope and networks of care, coupled with a deep fearlessness and revolutionary pragmatism – working within and outside, in mutually respectful ways, using ethical principles across urban terrain.

Embracing ethical realism, actions designed to shift and reframe political culture towards progressive values (Lakoff 2004) can be open, diverse and inclusive of a vast range of activists and protests groups, creative social entrepreneurs, and so on. Whether they are characterised more as deliberative ethical spectacles in the city or are undertaking the more mundane but necessary processing of resetting rules, testing new economic exchange, the ethical city provides minimal key touchpoints to guide associated endeavours. Boyd and Duncombe (2004) suggest actions that are more spectacular could be: participatory; open to shifting contexts and ideas; transparent in that they avoid tricking or deceiving participants; realistic in revealing real-world power dynamics and making visible the invisible and utopian in making the impossible, possible. The ethical city offers a mosaic of initiatives in cities that are intentional and recognisable due to the processes and conduct involved, and the nature of the focus upon tackling inequality and environmental concerns. As we have sought to show through this book, the challenge is to remain critical yet hopeful; reflexive yet engaged. Whether seemingly technical or economic (community wealth building, participatory budgeting, universal basic income etc.) or more overtly creative or activist protest (fearless cities, ethical spectacles etc.) a vast array of experimentation and change is relevant in redressing the injustices of contemporary cities.

References

Avelino, F. (2009) Empowerment and the challenge of applying transition management to ongoing projects, *Policy Sciences 42* (4): 369–390. https://doi.org/10.1007/s11077-009-9102-6.

Avelino, F., Grin, J., Pel, B. and Jhagroe, S. (2016) The politics of sustainability transitions, *Journal of Environmental Policy & Planning 18* (5): 557–567. https://doi.org/10.1080/1523908X.2016.1216782.

Avelino, F. and Wittmayer, J.M. (2015) Shifting power relations in sustainability transitions: A multi-actor perspective, *Journal of Environmental Policy & Planning*: 1–22. https://doi.org/10.1080/1523908X.2015.1112259.

Baeumler, A., Ijjasz-Vasquez, E. and Mehndiratta, S. (2012) *Sustainable Low-Carbon City Development in China, Directions in Development: Cities and Regions.* Washington: World Bank. [Online]. Available: http://documents.worldbank.org/curated/en/576131468261265617/pdf/672260PUB0EPI0067848B09780821389874.pdf [Last accessed: 11 May 2020].

Barcelona City Council (2016) *General Principles of Technological Sovereignty.* [Online]. Available: www.barcelona.cat/digitalstandards/en/tech-sovereignty/0.1/general-principles [Last accessed: 6 April 2020].

Barrett, B.F.D. (ed.) (2005) *Ecological Modernisation and Japan.* London: Routledge.

Bijker, W.E., Hughes, T.P., Pinch, T. and Douglas, D.G. (1987) *The Social Construction of Technological Systems: New Directions in the Sociology and History of Technology.* Cambridge: MIT Press.

Bloom, P. (2017) *The Ethics of Neoliberalism: The Business of Making Capitalism Moral.* London and New York: Routledge.

Boyd, A. and Duncombe, S. (2004) The manufacture of dissent: What the left can learn from Las Vegas, *Journal of Aesthetics and Protest 1* (3). [Online]. Available: http://andrewboyd.com/the-manufacture-of-dissent-what-the-left-can-learn-from-las-vegas/ [Last accessed: 11 May 2020].

Bregman, R. (2017) *Utopia for Realists and How We Can Get There.* London and New York: Bloomsbury.

Bulkeley, H., Coenen, L., Frantzeskaki, N., Hartman, C., Kronsell, A., Mai, L., Marvin, C., McCormick, K., van Steenbergen, F. and Voytenko Palgan, Y. (2016) Urban living labs: Governing sustainability transitions, *Current Opinion in Environmental Sustainability 22*: 13–17. https://doi.org/10.1016/j.cosust.2017.02.003.

Byrne, B., Taminiau, J., Kim, K.N., Seo, J. and Lee, J. (2016) A solar city strategy applied in six municipalities: Integrating market finance, and policy factors for infrastructure-scale photovoltaic development in Amsterdam, London, Munich, New York, Seoul and Tokyo, *WIREs Energy Environ 5*: 68–88. https://doi.org/10.1002/wene.182.

Calvino, I. (1972) *Invisible Cities.* London: Vintage Books, published in 1997.

Caprotti, F. and Cowley, R. (2016) Interrogating urban experiments, *Urban Geography*: 1–10. https://doi.org/10.1080/02723638.2016.1265870.

Centre for Cities (2018) *Cities Outlook 2018.* [Online]. Available: www.centreforcities. org/wp-content/uploads/2018/01/18-01-12-Final-Full-Cities-Outlook-2018.pdf [Last accessed: 24 March 2020].

Chen, J. (2017) *Future Job Automation to Hit Hardest in the Low Wage Metropolitan Areas Like Las Vegas, Orlando and Riverside-San Bernardino,* ISEA Publish. [Online]. Available: www.iseapublish.com/index.php/2017/05/03/future-job-automation-to-hit-hardest-in-low-wage-metropolitan-areas-like-las-vegas-orlando-and-riverside-san-bernardino/ [Last accessed: 6 April 2020].

City of Zurich (2011) *On the Way to the 2000-Watt Society: Zurich's Path to Sustainable Energy Uses.* [Online]. Available: www.stadt-zuerich.ch/portal/en/index/portraet_der_stadt_zuerich/2000-watt_society.html [Last accessed: 24 March 2020].

City of Zurich (2016) *Roadmap: 2000-Watt Society, Zurich.* [Online]. Available: www.stadt-zuerich.ch/content/dam/stzh/portal/English/portrait_city_of_zuerich/Documents/Report_Roadmap-2000-watt-society-einseitig.pdf [Last accessed: 24 March 2020].

Coffman, M., Wee, S., Bonham, C. and Salim, G. (2016) A policy analysis of Hawaii's solar tax credit, *Renewable Energy 85*: 1036–1043. https://doi.org/10.1016/j.renene.2015.07.061.

Conger, K., Fausset, R. and Kovaleski, S.F. (2019) San Francisco bans facial recognition technology, *New York Times.* [Online]. Available: www.nytimes.com/2019/05/14/us/facial-recognition-ban-san-francisco.html [Last accessed: 24 March 2020].

Costanza, R. (2000) Visions of alternative (unpredictable) futures and their use in policy analysis, *Conservation Ecology 4* (1). [Online]. Available: www.consecol.org/vol4/iss1/art5/ [Last accessed: 28 April 2020].

Cranshaw, J. (2013) *Whose "City of Tomorrow" is it? On urban computing, utopianism, and ethics,* UrbComp'13: The 2nd ACM SIGKDD International Workshop on Urban Computing, August 11–13, 2013, Chicago, IL. [Online]. Available: SSRN: https://ssrn.com/abstract=2289964 [Last accessed: 27 April 2020].

Crompton, T. and Weinstein, N. (2015) Common cause communication: A toolkit for charities, *Common Cause Foundation.* [Online]. Available: Common cause communication: A toolkit for charities [Last accessed: 8 May 2020].

de Boer, J. (2015) Resilience and the fragile city, *Stability, International Journal of Security and Development 4* (1). Art. 17: 1–7. http://doi.org/10.5334/sta.fk.

Duncombe, S. (2019) *Dream or Nightmare: Reimagining Politics in an Age of Fantasy.* New York and London: OR Books.

Elliott, C. (2019) Planning for a 'just transition': Leaving no worker behind in shifting to a low carbon future, *World Resources Institute.* [Online]. Available: www.wri.org/blog/2019/03/planning-just-transition-leaving-no-worker-behind-shifting-low-carbon-future [Last accessed: 11 May 2020].

Evans, J. and Karvonen, A. (2014) 'Give me a laboratory and I will lower your carbon footprint!': Urban laboratories and the governance of low-carbon futures, *International Journal of Urban and Regional Research 38* (2): 413–430. https://doi.org/10.1111/1468-2427.12077.

Fisher, M. (2009) *Capitalist Realism: Is There No Alternative?* Winchester and Washington: Zero Books and John Hunt Publishing.

FitzPatrick, W.J. (2015) Debunking evolutionary debunking of ethical realism, *Philosophy Studies 172*: 883–904. https://doi.org/10.1007/s11098-014-0295-y.

Frase, P. (2016) *Four Futures: Life after Capitalism.* London: Verso Books.

Fussey, P. and Murray, D. (2019) *Independent Report on the London Metropolitan Police Service's Trial of Live Facial Recognition Technology,* Human Rights Centre, University of Essex. [Online]. Available: www.hrbdt.ac.uk/download/independent-report-on-the-london-metropolitan-police-services-trial-of-live-facial-recognition-technology/# [Last accessed: 11 May 2020].

Geels, F.W. (2005) The dynamics of transitions in socio-technical systems: A multi-level analysis of the transition pathway from horse-drawn carriages to automobiles, (1860–1930), *Technical Analysis and Strategic Management 17* (4): 445–476. https://doi.org/10.1080/09537320500357319.

Geels, F.W. (2010) Ontologies, socio-technical transitions (to sustainability), and the multi-level perspective, *Research Policy 39* (4): 495–510. http://dx.doi.org/10.1016/j.respol.2010.01.022.

Geels, F.W. (2014) Regime resistance against low-carbon transitions: Introducing politics and power into the multi-level perspective, theory, *Culture & Society 31* (5): 21–40. https://doi.org/10.1177/0263276414531627.

Geels, F.W., Berkhout, F. and van Vuuren, D.P. (2016) Bridging analytical approaches for low-carbon transitions, *Nature Climate Change*. [Online]. Available: www.nature.com/doifinder/10.1038/nclimate2980 [Last accessed: 11 May 2020].

Geels, F.W. and Schot, J. (2007) Typology of sociotechnical transition pathways, *Research Policy 36*: 399–417. https://doi.org/10.1016/j.respol.2007.01.003.

Gellatly, J. and Rivero, M. (2018) Radical municipalism: Fearless cities, *P2P Foundation*. [Online]. Available: https://blog.p2pfoundation.net/radical-municipalism-fearless-cities/2018/04/03 [Last accessed: 25 March 2020].

Giddens, A. (1998) *The Third Way: The Renewal of Social Democracy*. Cambridge: Polity Press.

Giddens, A. (2000) *The Third Way and Its Critics*. Cambridge: Polity Press.

Gimenez, A.F.C. and Tamajon, L.G. (2019) An analysis of the process of adopting local digital currencies in support of sustainable development, *Sustainability 11* (849). https://doi.org/10.3390/su11030849.

Giridharadas, A. (2019) *Winner Takes All: The Elite Charade of Changing the World*. New York: Vintage.

Graham, S. (2010) *Cities under Siege: The New Military Urbanism*. London: Verso Books.

Hodson, M. and Marvin, S. (2010) Can cities shape socio-technical transitions and how would we know if they were? *Research Policy 39*: 477–485. https://doi.org/10.1016/j.respol.2010.01.020.

Holmgren, D. (2009) *Future Scenarios: How Communities Can Adapt to Peak Oil and Climate Change: Mapping the Cultural Implications of Peak Oil and Climate Change*. Totnes: Green Books.

Horne, R. (2018) *Housing Sustainability in Low Carbon Cities*. London: Routledge.

Huberlant, A. and VerKamp, A. (2018) *Fearless Cities: Brussels*. [Online]. Available: https://commonspolis.org/wp-content/uploads/2020/02/Fearless-cities_Report_VF.pdf [Last accessed: 25 March 2020].

Hunt, E. (2018) More than 100 cities now mostly powered by renewable energy, data shows, *The Guardian*. [Online]. Available: www.theguardian.com/cities/2018/feb/27/cities-powered-clean-energy-renewable [Last accessed: 27 April 2020].

ICLEI (2019) *Urban Transitions Alliance Roadmaps: Sustainability Transition Pathways from Industrial Legacy Cities*. [Online]. Available: http://e-lib.iclei.org/wp-content/uploads/2019/04/Urban-Transitions-Alliance-Roadmaps-April-2019-web.pdf [Last accessed: 11 May 2020].

Inayatullah, S. and Black, P. (2020) Neither a black swan nor a zombie apocalypse: The futures of a world with the covid-19 coronavirus, *Journal of Futures Studies*. [Online]. Available: https://jfsdigital.org/2020/03/18/neither-a-black-swan-nor-a-zombie-apocalypse-the-futures-of-a-world-with-the-covid-19-coronavirus/ [Last accessed: 15 April 2020].

Institute for Global Environmental Strategies (2019) *1.5-Degree Lifestyles: Targets and Options for Reducing Lifestyle Carbon Footprints: Technical Report*. [Online]. Available: www.iges.or.jp/en/publication_documents/pub/technicalreport/en/6719/15_ Degree_Lifestyles_MainReport.pdf [Last accessed: 11 May 2020].

Jackson, T. (2009) *Prosperity without Growth: Foundations for the Economy of Tomorrow*. London and New York: Routledge.

Kissler, S.M., Tedijanto, C., Goldstein, E., Grad, Y.H. and Lipsitch, M. (2020) Projecting the transmission dynamics of SARS-Cov-2 through the post-pandemic period, *Science*. [Online]. Available: https://science.sciencemag.org/content/early/2020/04/14/ science.abb5793 [Last accessed: 15 April 2020].

Kitchin, R. (2019) The ethics of smart cities, *Raidió Teilifís Éireann*. [Online]. Available: www.rte.ie/brainstorm/2019/0425/1045602-the-ethics-of-smart-cities/ [Last accessed: 10 May 2020].

Klein, N. (2007) *The Shock Doctrine: The Rise of Disaster Capitalism*. New York: Metropolitan Books.

Ko, Y., Barrett, B.F.D., Copping, A.E., Sharifi, A., Yarime, M. and Wang, X. (2019) Energy transitions towards local carbon resilience: Evaluation of disaster-triggered local and regional cases, *Sustainability 11* (6801): 1–23. https://doi.org/10.3390/su11236801.

Kolbert, E. (2015) Can climate change cure capitalism?: An exchange, *The New York Review of Books*. [Online]. Available: www.nybooks.com/articles/2015/01/08/can-climate-change-cure-capitalism-exchange/ [Last accessed: 28 April 2020].

Kunstler, J.H. (2005) *The Long Emergency: Surviving the End of Oil, Climate Change, and Converging Catastrophes of the Twenty-First Century*. New York: Grove Press.

Lakoff, G. (2004) *Don't Think of an Elephant! Know Your Values and Frame the Debate*. Hartford: Chelsea Green Publishing.

Loorbach, D. (2007) *Transition Management: New Mode of Governance for Sustainable Development*. Utrecht: International Books.

Luque-Ayala, A., Marvin, S. and Bulkeley, H. (2018) *Rethinking Urban Transitions: Politics in the Low Carbon City*. London: Routledge.

Magrini, E. and Clayton, N. (2018) Can cities outsmart the robots? The future of skills in the UK cities, *Centre for Cities*. [Online]. Available: www.centreforcities.org/wp-content/uploads/2018/10/2018-10-12-Can-cities-outsmart-the-robots-The-future-of-skills-in-UK-cities-2.pdf [Last accessed: 6 May 2020].

Meadowcroft, J. (2011) Engaging with the politics of sustainability transitions, *Environmental Innovation and Societal Transitions 1* (1): 70–75. http://dx.doi.org/10.1016/j. eist.2011.02.003.

Metcalf, T. (2015) Ethical realism, 1,000 word philosophy: An introductory anthology. [Online]. Available: https://1000wordphilosophy.com/2015/11/05/ethical-realism/ [Last accessed: 6 April 2020].

Miller, H. and Asare, I.N. (2018) Why every city needs to take action on AI, *Oxford Insights*. [Online]. Available: www.oxfordinsights.com/insights/2018/8/1/why-every-city-needs-to-take-action-on-ai [Last accessed: 28 April 2020].

Mol, A. and Spaargaren, G. (2000) Ecological modernisation theory in debate: A review, *Environmental Politics 9*: 17–49.

Moore, T., de Haan, F., Horne, R. and Gleeson, B. (eds.) (2018) *Urban Sustainability Transitions: Australian Cases, International Perspectives*. Springer Series: Theory and Practice of Urban Sustainability Transitions. Tokyo/Berlin: Springer.

Muggah, R. (2016) *How fragile are our cities?* Global Agenda, World Economic Forum. [Online]. Available: http://www.weforum.org/agenda/2016/02/how-fragile-are-our-cities [Last accessed: 26 April 2020].

Nevens, F., Frantzeskaki, N., Gorissen, L. and Loorbach, D. (2013) Urban transition labs: Co-creating transformative action for sustainable cities, *Journal of Cleaner Production 50*: 111–122. https://doi.org/10.1016/j.jclepro.2012.12.001.

Newton, P. and Bau, X. (2008) Introduction: Transitions: Pathways towards sustainable urban development, chapter 1 in Newton, P. (ed.) *Transitions: Pathways towards Sustainable Urban Development*. Clayton: CSIRO Publishers.

Notter, D.A., Meyer, R. and Althaus, H.-J. (2013) The Western lifestyle and its long way to sustainability, *Environmental Science & Technology 47* (9): 4014–4021. [Online]. Available: http://pubs.acs.org/doi/abs/10.1021/es3037548 [Last accessed: 11 May 2020].

Pitroda, S. and Miailhe, N. (2017) Introduction: The rise of AI & robotics in the city, *Journal of Field Actions*. [Online]. Available: http://journals.openedition.org/factsreports/4377 [Last accessed: 22 December 2018].

Ravetz, J. (2020) *Pandemic 3.0: From Crisis to Transformation*. [Online]. Available: https://urban3.net [Last accessed: 15 April 2020].

Russell, B. (2019) Fearless cities municipalism: Experiments in autogestion, *Open Democracy*. [Online]. Available: www.opendemocracy.net/en/can-europe-make-it/fearless-cities-municipalism-experiments-in-autogestion/ [Last accessed: 17 March 2020].

Solecki, W., Grimm, N., Marcotullio, P., Boone, C., Bruns, A., Lobo, J., Luque, A., Romero-Lankao, P., Young, A. and Zimmerman, R. (2019) Extreme events and climate adaptation-mitigation linkages: Understanding low-carbon transitions in the era of global urbanization Wiley interdisciplinary reviews, *Climate Change 10*: e616. https://doi.org/10.1002/wcc.616.

Srnicek, N. and Williams, A. (2015) *Inventing the Future: Postcapitalism and a World without Work*. London: Verso Books.

Taminiau, J., Seo, J. and Lee, J. (20150 Energy dilemma of ethical cities and the solar city's promise. [Online]. Available: http://freefutures.org/energy-dilemma-of-ethical-cities-and-the-solar-citys-promise/ [Last accessed: 10 May 2020].

Thomas, K.E. (2017) Pittsburgh mayor wants a road map to ethical driverless cities, *Next City*. [Online]. Available: https://nextcity.org/daily/entry/driverless-cars-pittsburgh-uber-testing-city-asks-the-company-to-give-back [Last accessed: 24 March 2020].

Thomson, A., Lanxon, N. and Bloomberg (2019) Uber's London ban may just be the beginning of a global ride hailing backlash, *Fortune*. [Online]. Available: https://fortune.com/2019/11/27/uber-london-ban-global-ride-hailing-backlash/ [Last accessed: 24 March 2020].

Tun, Z.T. (2020) Top cities where Airbnb is legal or illegal, *Investopedia*. [Online]. Available: www.investopedia.com/articles/investing/083115/top-cities-where-airbnb-legal-or-illegal.asp [Last accessed: 24 March 2020].

Van den Bergh, J.C.J.M., Truffer, B. and Kallis, G. (2011) Environmental innovation and societal transitions: Introduction and overview, *Environmental Innovations and Transitions 1* (1): 1–23. https://doi.org/10.1016/j.eist.2011.04.010.

Vincent, J. (2020) London police to deploy facial recognition cameras across the city, *The Verge*. [Online]. Available: www.theverge.com/2020/1/24/21079919/facial-recognition-london-cctv-camera-deployment [Last accessed: 24 February 2020].

Voytenko, Y., McCormick, K., Evans, J. and Schliwa, G. (2016) Urban living labs for sustainability and low carbon cities in Europe: Towards a research agenda, *Journal of Cleaner Production 123*: 45–54. https://doi.org/10.1016/j.jclepro.2015.08.053.

Werner, R. (1983) Ethical realism, *Ethics 93* (4): 653–679. https://doi.org/10.1086/292487.

ANNEX 1

Issues covered in the city scan pilot survey

Category	Description	Topics included (22)	Issues (157)
City Development	Relates to challenges and actions around topics such as: labour and human rights; intra-generational equity; equal opportunity; universal access to resources and justice; sustainable economic development	Social Inclusion and Equality Education Health and Well-being Social Care Services Food Security Access to Adequate Housing Access to Mobility and Public Transport Community and Cultural Identity Access to Employment Fair Work Public Safety and Security City/Region Security	Participation of indigenous peoples and minority ethnic groups Women's rights Discrimination based on age, race, religion, gender, sexual preferences, ethnicity, etc. Poverty Access to education for children (early education up to 5 years old) Access to higher education (vocational training, university, etc.) Access to education for adults and elderly Quality of education Access to free education/high cost of education Literacy Access to post-primary education Access to primary education Access to adequate healthcare Nutritional health of citizens Substance abuse Access and provision of health insurance Access to sports facilities and/or programmes Access to dental care

Access and/or availability of preventative care initiatives
Maternal and child health services
Access to social care services
Provision of social care services
Access to and/or provision of social security
Access to nutritional food
Cost of food
Nutrition and related diseases (undernourishment, obesity, diabetes, etc.)
Availability of food
Seasonal variability and/or access to food
Access to adequate housing (housing quality)
Access to adequate social or public housing
Housing availability (housing stock)
Housing affordability
Access to housing support services and infrastructure (energy, water, etc.)
Informal settlements
Land tenure
Availability of appropriate land for housing
Quality of public transport
Cost of public transport
Road quality
Cleaner transport options
Safe passage for bikes
Safe pedestrian mobility
Public transport connectivity
Community identity
Support for cultural activities
Infrastructure and resources for cultural activities
Tolerance of cultural differences
Inclusion of cultural identities

(*Continued*)

(Continued)

Category	Description	Topics included (22)	Issues (157)
			Freedom to express cultural diversity
			Artistic expression in the city/region
			Protection of heritage
			Respect for arts and heritage
			Safe and favourable working conditions
			Unemployment
			Access for women to employment opportunities
			Access to maternity benefits
			Access for marginalised and/or other ethnic groups to employment opportunities
			Access to fair work
			Access to local employment
			Operational and/or business management issues
			Process to establish local enterprises
			Restrictions/freedom to join workers associations or trade unions
			Activity of trade unions
			Compulsory and forced labour
			Child labour
			Equal and fair pay
			Discrimination based on age, gender, sexual preference, religion, ethnicity, etc.
			Persistent and/or increasing poverty
			Migration (i.e. outgoing, incoming, uncontrolled migration)
			Crime
			Perception of safety (e.g. sense of danger or risk)
			Perception of security (e.g. lack of support networks)
			Disaster risk management systems
			Vulnerability to disasters (natural and human induced)
			Capacity to respond to disasters
			Terrorism
			Warfare
			Civil uprising and/or civil unrest

City Sustainability	Relates to challenges and actions around topics such as: environment; intergenerational equity; resilience	Environmental Sustainability	Sustainable practices in industry
		Water Management	Access to public green spaces
		Energy Security	Natural resource management
		Waste Management and Reduction	Environmental pollution
		Climate Change Mitigation	Deforestation and land clearing
		Climate Change Adaptation	Air pollution
			Ecosystems at risk (e.g. surrounding mangroves, grasslands, etc.)
			Land management
			Street tree coverage
			Access to safe and potable water for all citizens
			Access to sufficient potable water
			Water pollution
			Water recycling and reuse
			Sewerage infrastructure
			Sewerage treatment and disposal
			Localised or distributed water capture
			Street drainage
			Consistent access to energy for all citizens
			Energy scarcity
			Cost of energy
			Clean energy alternatives
			Dependency on non-renewable / fossil fuel energy
			Street lighting coverage
			Illegal access to energy supply network
			Localised or distributed energy options
			Energy efficient alternatives (e.g. light bulbs, appliances, buildings)
			Hazardous commercial and industrial waste
			Levels of recycling
			Litter

(Continued)

(Continued)

Category	Description	Topics included (22)	Issues (157)
			Waste management infrastructure
			Solid waste collection and management
			Unregulated burning of waste
			Appropriate land for waste disposal
			Greenhouse gas (carbon) emissions from industry
			Greenhouse gas (carbon) emissions from transport
			Greenhouse gas (carbon) emissions from housing
			Greenhouse gas (carbon) emissions from other buildings
			Greenhouse gas (carbon) emissions from other sources
			Greenhouse gas (carbon) emissions from city organisational operations
			Greenhouse gas (carbon) emissions from waste
			Greenhouse gas (carbon) emissions from land and agriculture
			Community understanding of climate change issue
			Severe and/or prolonged droughts
			Increased severity and/or frequency of extreme weather events
			Slow onset impacts of climate change (i.e. slowly changing weather patterns: precipitation patterns, changing seasonal temperatures)
			Sea-level rise
			Flooding
			Landslides and/or unstable ground
			Adequacy of infrastructure to deal with likely future impacts
			Community capacity to prepare for and respond to above events
			City staff capacity to prepare for and respond to above events

City Governance	Relates to challenges and actions around topics such as: anti-corruption; leadership; transparency and accountability; and participation.	Transparency and Accountability	Public access to city/region government information
			Accountability of city leaders
			Transparency of processes
			Conflict of interest
			Accountability of bureaucrats
			Transparent procurement processes
			Fair regulation enforcement
			Public reporting against urban objectives
		Community Participation	Capacity and skills within the community to engage with the local government
			Interest from the community in engaging with the local/regional government
			Trust in local/regional government
			Recognised and/or formalised city processes of community engagement
			Processes where community engagement can effect change
		Resources and Leadership	Financial resources
			Capacity to secure external funding
			Leadership and management skills
			Regional Internet provision
			Modern IT infrastructure
			Adequate staff skills
			Strength of community institutions
			Strength of private/business sector
			State or national support
			Adequate and appropriate remuneration of public sector staff
			Appropriate number of staff
		Anti-corruption in City and Political Processes	Political corruption
			Bureaucratic corruption
			Corruption in the private sector
			Arbitrary and impartial judicial processes

INDEX

Note: Page numbers in *italic* indicate a figure and page numbers in **bold** indicate a table on the corresponding page.

1948 Universal Declaration of Human Rights 24, 62
2008 Global Financial Crisis (GFC) 5–6, 34, 51, 75, 150, 164; aftermath of 5–6, 34, 125; public sentiment after 75
2015 One New York City (One NYC) plan 48
2016 [Oslo] Climate and Energy Strategy 156
2018 Social Progress Index 127
2019 Carbon-Roadmap for the City of Leeds 154
2030 Agenda for Sustainable Development

accelerationism 126, 128
activism 7, 33, 65, 125; civic 11; local 149
affordable housing 48, 50–51, 66, 97, 113, 146
Agenda 21 86–87
agonistic exchanges 48
agonistic urbanism 8
Agyeman, Julian 28
AI *see* artificial intelligence
Airbnb 177
Airtasker type apps 177
alliances, building 179–180
Alpha Cities *see* competitive city orientation

Ankara Metropolitan Municipality, women's resistance to 32–33
anti-corruption 43, 93, 158
anti-elite populism 76
anti-system politics 4, 76
Anywhere but Westminster (video series) 4
Arnstein, Sherry 72
artificial intelligence 129, 132; adoption of 173; ethical concerns raised in relation to 174; fears abound around 179; investments in 129, 132; market size pre-Covid-19 129
Asian Ethical Urbanism, notion of 16
austerity 4, 5, 62, 65; and inequality 14; post-GFC 125
Australia: research on ethical practice in 43; urban liveability assessment of cities 110; work activities to be automated in 129
Austria, social housing in 50
automation 41, 125, 135, 151, 173–174; adoption of 173; and artificial intelligence 129, 132; capital and productivity benefits of 132; fears abound around 179; impact on job 128–129, 174; industrial robots 128; potential by Australian local government areas 129, *131*; potential by state in United States 129, *130*; ready countries 132; strategies for benefiting from 132
Autonomist Municipalism 147, **148**

Barber, Benjamin 66
Barcelona: affordable housing crisis 51; ECS Results 2015 94, *96*; technological sovereignty implemented in 136
Barcelona en Comú (Barcelona in Common): code of political ethics 62–63, **64**; strategies for change 65
Barcelona Minimum-Income project 152, **153**
basic values, conceptualisation of 10, *12*; *see also* values, disconnect between cities and
'BAU' scenario of entrenched deregulated capitalism 173
behaviour change for transition 175
Berlin, carbon emission reduction by 48, 154–155, **156**
Besos (neighbourbood in Barcelona) 152
Big Leeds Climate Conversation 47
Big Tech Economy 135
Blade Runner (film) 124
Bloomberg, New York Mayor 27
B-MINCOME project *see* Barcelona Minimum-Income project
Boston, ethnic minority struggle for justice in 28
bribery 53, 77
Bristol: introduction of local currencies 150–151; local decarbonisation efforts 157; localisation of goals by 90; VLR process 90, 92
Bristol City Council 90
Bristol Green Capital Partnership 90, 92
Bristol One City Plan, cross reference local policies in 92
Bristol Pound (£B) 150–151, 177
Brown, Matthew 147
Brown, Wendy 4, 42
Bunge, Mario 26–27
business-unusual solutions and disruption 135–138

C40 cities initiative 133, 135
Cairo, gender-based harassment and violence in 32
Calvino, Italo 1, 18–20, 24, 41, 61, 86, 104, 124, 144, 164
Campanella, Tommaso 177
capitalism: 'BAU' scenario of entrenched deregulated 173; contemporary late 4; overcoming fear of end of 178–179; surveillance 125
capitalist realism 7, 164–165
carbon emission reduction 48, 52, 79, 132, 154; Bristol One City Plan 157–158;

C40 cities initiative 133, 135; City of Berlin 154–155, 156, **156**; City of Oslo 156–157; City of Zurich 175, *176*; integration with spatial planning 154; Leeds Carbon-Roadmap 154, *155*; role of communities in 175; urban living labs emerging from 170
carbon emissions: consumption-related, increase in 132; measurement challenges 135; production-related 132
carbon intensity and income inequality 99
carbon neutrality 48, 52, 133, 157
carbon sinks 133
'carnival communities' *see* 'explosive communities'
Castells, Manuel 19, 24, 29, 106, 125
casualisation of work 126
categorical imperative 45, 51–52, 180
Centre for Health, Environment and Justice 28
Centre for Local Economic Strategies 74
Chadwick, Stephanie 'Steve' 65–66
Chan, Jeffrey K.H. 16
Chinese AI industry 129
cities: agency 149; assessment criteria 28; autonomy 66; cross-community mobility in 10; economic development 79; as intersections of ideas 18; life in 14; as nodes 11; overarching challenges of 9; rankings 106, 108, 110, **110**; as sanctuaries for vulnerable groups 13; *see also* ethical cities; fearless cities
Citizen's Basic Income Initiative 153
city governments: as facilitators 79; integrating SDGs within practices 86; local voluntary reviews (LVRs) 87–88
city-network initiatives 89
city shaping *see* ethical city shaping
City We Need 2.0, The (document) 30
civic engagement strategies 100
civil society organisations 30
CLES *see* Centre for Local Economic Strategies
Cleveland Evergreen Initiative 146
Cleveland Model 145–146
climate change 9, 13–14, 25, 34–36, 48, 136, 138, 149, 154, 166, 169, 172; Google web search for 6; overcoming fear of 178; policy and practice, disconnects between 13
climate crisis 2, 7, 35, 71, 144, 154
climate responses 9
climate science 47
Colau, Ada 65
collaborative consumption 79, 149, 177

collective action 5, 34, 36
collective endeavours 6–7
collectivist approaches 177
Common Cause Foundation 11
Commons housing development
    project 52
community: activism 33; austerity and
    Brexit impact on 4; communitarian 4;
    definition of 1; elusive 4; as goal 11;
    inclusiveness 2; new forms of 4
community-based organisations 32, 33
community leadership, public trust in
    77–78
community self-provisioning 175
community wealth building 63, 74,
    145–146, 148–149, 158, 165, 167
comparative ethics *see* descriptive ethics
competition 61, 108, 110
competitive advantage 105
competitive city agendas 113
competitive city orientation 105,
    106–110; competitive dynamos for
    global economy 106; facets of 108;
    unavoidable consequences of 108;
    see also global cities
consequentialism 44–46, 49–50, 68
contemporary city, ethics in 42–44
contemporary collective action 35
contemporary politics and leadership 75, 76
contemporary populist political
    engagement 80
contemporary urbanism 8, 108
Copenhagen, emissions in 135
corruption 2, 64, 73, 75, 77, 88,
    93–94, 97, 100, 110, 114–115,
    117–118, 145, 178
Covid-19 pandemic 6, 49, 108, 127, 129,
    164, 180
CPTED *see* Crime Prevention Through
    Environmental Design
Crime Prevention Through
    Environmental Design 98
critical urban theory 15, 24, 25, 29, 34
cross-community mobility, in cities 10
crowdsourcing 32
cultural relativism 52–53, 56

Davis, Mike 111
decarbonisation of local economies
    154–158; Bristol One City Plan
    157–158; carbon emission cuts 154;
    City of Berlin 154–155, 156, **156**; City
    of Oslo 156–157; City of Zurich 175,
    176; Leeds Carbon-Roadmap 154, *155*;
    linked to Paris Agreement 157; new

technologies for 171; transition away
    from fossil fuels 174; UK cities 133
decisions, in ethical city 41, 52
Democracy Collaborative 74–75, 79
democratic glocalism 66
'Demolition Office' 32
demos 1, 4, 147, 167, 178, 180
deontological ethics 50, 63, 74, 158;
    component parts of 50; housing 50–51
deontology 44–46, 49–50, 56
descriptive ethics 53
destructive entrepreneurialism 73
DFG Research Group on Urban Ethics 17
Dhaka, global ranking of 110
digital monetary systems 152, 153, 177
digital platforms, negative impacts
    associated with 41
digital social currency *see* digital
    monetary systems
Dikmen Valley neighbourhood 32–33
Diprose, Rosalyn 3
disruption and business-unusual solutions
    135–138
disruptive smart city technologies, ethical
    implications of 41
distressed economy 127
distrust in politics 76–77; *see also* public
    trust in government
Droit à la ville, Le (Lefebvre) 19, 24
Duto, Olivia 72
dystopian cities 124–126
dystopic storytelling 124, 125

economic growth, conventional measures
    of 174–175
Economist Intelligence Unit (EIU),
    Global Liveability Index 108, **110**
ECS *see* Ethical City Scan
Edelman Trust Barometer 75; post-GFC
    75; public trust in local government 78
Egypt, gender-based harassment and
    violence in 32
elitist visions 125
emotivism 56
Empathy Economy 135, **137**
employment apocalypse 128
employment opportunities *see* jobs
energy production, local control of 48
enfranchising of working men 24
entrepreneurship: Democracy
    Collaborative 74–75; deontological
    dimensions of ethics 74; groups
    73; implications for profits 73–74;
    Mondragon model 74; normative
    ideas of 73

environmental ethics 1, 3
environmental justice 28
environmentally sustainable economies 136
environmental problems, societal
    responses to 5
ETB *see* Edelman Trust Barometer
Ethical Cities MOOC 53–54, *55*, 62, 93,
    94, 97, 98
ethical citizenship 71
ethical cities 57; assessment 98–100;
    challenges for 5; characteristics of 2;
    cognizant and responsive 171–172;
    as evolving narrative 7; concept 2, 3,
    168; core urban concerns addressed
    by 2–3; definitions of 1, 2, 5, 15, 47;
    development 100; diagnostic tool 92;
    engagement around SDGs, need for
    100–101; experiments 78–79; goals
    of 1; inclusiveness and universality 1;
    measuring progress towards 92–97;
    origins of 14–17; as process 177–178;
    progress in 9; reimagining 158–159;
    rights to 34–36; urban transitions
    to 172–173; *vs.* neoliberal city 7–8;
    see also ethical city shaping; fragile city
    orientation; ontological orientations;
    right to the city; values, disconnect
    between cities and
Ethical City Scan: Barcelona ECS
    Results 2015 94, *96*; cities surveyed in
    93–94, *95*; concerns emerging from
    97, *98*; development of 92; indicator
    framework **93**; methodology 92, 93;
    ranking results of top ten issues 94, **97**;
    strength of 94
ethical city shaping: challenges
    confronting 75–78; conflicting interests
    62, **63**; ethical leadership 62–67; factors
    influencing 61; outcomes of 80;
    prospects for 78–80; Rotorua, New
    Zealand 65–66
ethical communities 100
ethical community entrepreneurs 73;
    Democracy Collaborative 74–75;
    Mondragon Corporation 74
ethical democracy 178
ethical digital standards 172
ethical dilemmas 8, 47
ethical evolutions 10
ethical foundations, rebuilding on 44, 47–48
ethical frames 56–57
ethical innovations 10
ethical investors 3
ethical leadership 62; Barcelona en
    Comú code of political ethics 62–63,

64; characteristics of 66–67; city
    administrations and 62; ethical
    conversations and reflections for
    101; and governance 77; limits and
    opportunities for 65, **65**; political
    representatives 62, 63; Rotorua, New
    Zealand 65–66; *see also* Barcelona en
    Comú (Barcelona in Common)
ethical lens 19
ethical living, contemporary urban
    settings 17
Ethical Reading 16
ethical realism 164–168, *167*, 180
ethical subjects 7
ethical/unethical perceptions 78, **78**
ethical urban approaches 2, 9–10;
    behaviours 53; development 92, 149;
    futures 41–42
ethical urbanism 15, 47, 67–72, 164–165,
    167, 180
ethical urbanists 63, 67
ethical urbanites 62, 63
ethics 57; in contemporary city 42–44;
    definition 3; importance of 3–4
European cities, low turnouts in local
    elections of 77
evolutionary economics 170
exchange modes, strategies reconfiguring
    177
Exodus Economy 135, **137**
'explosive communities' 4–5
extrinsic values 11

facial recognition technology 173
Fainstein, Susan 27–29, 71, 108
fearless cities 39, 119, 145, 163, 166
Fearless Cities Summit, in Brussels 178
fearlessness 119–120, 180
feminisation of politics 145
feminist intrinsic values 178–179
feminist street art against street
    harassment 32
feminist urbanism 28
feral city: characteristics of **116**, 116–117;
    diagnostic tool for evaluating 117, **118**
fightback cities 105, 119–120
Fisher, Mark 7, 164
Fitzgerald: The Geography of a Revolution
    (Bunge) 26
fragile city orientation 105, 115–119,
    166; feral city **116**, 116–117; focus
    on addressing 117; insecurity 115;
    Medellín case 117, 119; pejorative
    adjectives applied to 115; ranking
    of most violent cities 115; rupture

in social contract 115; turbo-urbanisation 117
fragile economy 164
Fragile States Index 115–116

gated communities 17, 114
Gaviria, Aníbal 117
gentrification 27–28, 67, 98, 108, 146, 177
Glaeser, Edward 10
global cities 42, 89, 106–108, 114; characteristics of 106, 108; employment structure and functioning of 108, *109*; 2010 GaWC map 106, *107*; top ten ranked cities, in 2019 **110**; *see also* competitive city orientation
Global Compact Cities Programme 46, 66, 92–93, **93**, 95, 96, 98, 100
global economic competition 105
Globalisation and World Cities (GaWC) ranking 106
Global Municipalism 145
Global South 9, 115, 132, 153
Global Taskforce of Local and Regional Governments 89
governance 8, 17; inclusionary form of 9–10; integrity of institutions of 75–78; neoliberal 61; in sharing economies 79; *see also* public trust in government
Graham, Stephan 114
greenhouse gas (GHG) emission 48, 97, 135, 157
Green New Deal **46**, 136
'green *vs.* jobs' dichotomy 14

Habitat International Coalition 29
happy cities 18
HarassMap 32
Harvey, David 8, 16, 19, 24–25, 27, 29, 33–36, 71, 119, 125
healthcare access, elimination of inequities in 48–49
High-Level Political Forum (HLPF) on Sustainable Development, 87, 89, 90
homeless ghettos 51–52
Hopkin, Jonathan 4
horizontalism 66
housing development projects 52
human rights 2–3, 9, 24–25, 29–30, 41, 50, 61–63, 72, 88, 93, 116, 118, 135, 145, 149, 178

Ignatieff, Michael 8, 11
inclusive local economies: GVA leaks 150; introduction of local currencies 150–151; Minamata 150; participatory

budgeting 72, 153; universal basic income experiments 151–153
individualism 4, 6–7, 168
innovations 10, 78, 149, 158, 164, 170–171
insecurity 4, 10, 97, 115, 127, 145, 152, 169
institutional ethics 42
integrated city planning 157
interaction, nested scales of 5
intergenerational equity 113
Intergovernmental Panel on Climate Change: calls for action 132; 2018 Special Report 133
intrinsic values 10, 11, 17, 50, 57, 177, 178
intuitive ethics 56
Inventing the Future – Postcapitalism and a World Without Work (Srnicek and Williams) 179
Invisible Cities (Calvino) 1, 18, 24, 41, 61, 86, 124, 144, 164
IPCC *see* Intergovernmental Panel on Climate Change
ISO 37120 89

Jacobs, Jane 10, 15, 67
Japan 11, 52, 111, 126, 132, 150
Jeffko, Walter 3
jobs: automation impact on 128–129, 174; insecurity 127; manufacturing, decline in 126–127; social impact of loss of 127
judgements and justifications 53
just city 16, 27, 46

Kantian ethics 8
Kantian framing 50–51
Khan, Kublai 18
knowledge sharing 100, 173, 179–180

LA21 process *see* Local Agenda 21 process
labour market, scenarios in 2035 135–137, **137**
Ladder of Public Participation 72, *73*
Land, Nick 128
late capitalism 4, 19, 165, 168, 170
leadership *see* ethical leadership
least liveable cities 110
Leeds: Big Leeds Climate Conversation 47; CO₂ emission cuts 154
Lefebvre, Henri 19, 22, 24–26, 29–30, 29–31, 35–36, 38–39, 148
liberal democracies 76, 119
liveable city orientation 105, 110–114; of Australian city 110; inequality between cities 111; Londoners' concerns 113–114; measurement tools 111, 113;

poverty rates 111, *112*; security zones and 114
Local Agenda 21 process 87–88, 90, 100
local council, public trust in 78
local currencies 79, 150–151, 158, 177
local economies: development 79; initiatives 74, 146; stewardship 147
local elections, low turnouts in 77
local government 8; Denver, study of 43; ethic codes and guidelines 50
Local Governments for Sustainability 88
local neighbourhood revitalisation initiatives 145
local political processes, areas for renewal of 80
local voluntary reviews 87
London 77, 106, 108, 113, 114, 119, 173, 177
Londoners, concerns of 113–114
London Sustainable Development Commission 113
Los Angeles: code 43; *Ethics Handbook for City Official* 43; portrayed in *Blade Runner* (film) 124
Love Canal tragedy 28
low-carbon cities 133
LVRs *see* local voluntary reviews

Managed Municipalism 147, **148**
Māori people of Te Arawa 65
Marcuse, Peter 20, 24–25, 27, 29, 31, 33
mass tourism, response to 18
Medellín, transformation of 117, 119, participatory budgeting 154
mega-trends: accelerationism 128; casualisation of work 126; in late twentieth century 127; loss of manufacturing jobs 126–127; positive implications for planet 126; rapid innovation 127–128; social impact of job loss 127; technological change 127
Melbourne 19, 52, 110, 126
Mercer Quality of Living Ranking 108, **110**
meta-ethics 44, 45, 53, 54, 56
Metropolitan Revolution 79, 146
Mexico City, downtown streets in 27
migrants, global cities as destination points for 108
Millennium Development Goals 86–87
MIS *see* Municipal Inclusion Support
MLP *see* Multi-Level Perspective
Mondragon Corporation 74, 147, 148
money circulation patterns *see* regional economic cycle analysis

Monocle Quality of Life Survey 108, **110**
MOOC on ethical cities *see* ethical cities MOOC
moral codes 8
moral operating system 8, 11, 16
morality 18, 52, 53, 54
moral values and ethical alignment **15**, 15–16
Moses, Robert 67
Müller, Michael 48
Multi-Level Perspective 170
Municipal Council Agreement 154
municipal democracy 49
Municipal Inclusion Support 152
Municipal Socialism 145, 148

nationalist populism 76–77
national voluntary reviews 87
neighbourhood 28, 33, 67, 99, 146, 178
Nemeth, Jeremy 114
neoliberal cities 25, 42; local government in 8; *vs.* ethical cities 7–8
neoliberal urbanism 43
neoliberalism: austerity 148; core goal of 43–44; ethical reasoning 42; governance 61; impact on society 4; impact on urban planning policies 67, **68–70**, 71; individualizes ethics 6; from normative ethics perspective 43; orientation of 6; policies in capitalist city 33; unbridled 42
New Military Urbanism (Graham) 114, 166
new municipalism 8, 33, 145–150; in Baltimore 146; community wealth building 145–147; fearless cities 145; future realities for 149; innovations and 149; local economies initiatives 146; local neighbourhood revitalisation initiatives 145; scalar strategy of transformation 150; tripartite typology for conceptualising of 147, **148**, 149
New Sliding Scale of Ethics 54, 56
New Urban Agenda (NUA) 30
New York 27, 48, 89, 106, 114, 157
Nightingale housing development project 52
non-consequentialism *see* deontological ethics
non-normative ethics 15, 48; definition of 44; descriptive ethics 54, *55*; identification of influences by 53–54; meta-ethics 54, 56; sub-categories of 44, **45–46**, 53
normative ethics: consequentialism 49–50; deontological ethics 50–52;

focus on human conduct 44; moral obligation for existential threats 48–49; neoliberalism and 43; questions addressed by 48; sub-categories of 44, **45–46**; virtue ethics 52–53
Northern Powerhouses Initiative 146
Norton, Richard 116–117
NVRs *see* national voluntary reviews

Obama, Barak 75, 114
Økern Centre Living Lab 156
One New York: The Plan for a Strong and Just City 157
ontological orientations 104; competitive city orientation 105, 106–110; fightback cities 105, 119–120; fragile city orientation 105, 115–119; liveable city orientation 105, 110–114
Osaka 52, 110, 126
Oslo 79; ban on private cars 48; carbon emission cuts 156–157; social and environmental projects 79
Oslo City Council 156

pandemics 6–7, 9, 11, 13, 36, 48–49, 54, 71, 78, 108, 113, 149, 151, 164, 166, 169, 178
participatory budgeting 49, 72, 117, 148, 153–154, 158
people-centred approaches, to economic development 74
petty corruption 77
photovoltaic renewables transitions 171
place-based community development strategies 28
placelessness 27
Platform Municipalism 147, **148**
police: certification of buildings 98; militarisation, public concerns about 114
policy innovation 7
political mobilisation 34
politics of proximity 8, 147
Polo, Marco 18
populism 76–77, 171; anti-multicultural 165; nationalist 76; rhetoric from political leaders 76
Porto Alegre, city of 49, 72, 97
post-Brexit London, realities of 113
post-disaster governance 6
Potsdam Institute for Climate Impact Research 156
poverty rates 111, *112*
powersharing, pathways for 5
Prato, city of 16
precariat 127

Precision Economy 135, **137**
preference utilitarianism 45, 47, 49, 72, 158
Preston Model 146–147
private development 27
private sector investment 79
private spaces 27
production and consumption, interventions to 169
productive entrepreneurship 73
professional ethics 42, 43
property rights of owners 25
protest movements 7, 17
public money 5
public sentiment *see* public trust in government
public service 62
public spaces of children 26
public trust in government 76–77; democratic dissatisfaction 76; ETB results 75; local leadership in Australia 77–78; low turnouts in local elections 77; populism and 77; in United States 76; *see also* distrust in politics
PV systems 171

racial discrimination 27
radical social-cultural transformations 149
rebel cities 119
reflexivity 53
regional economic cycle analysis 150
renewable energy 157, 170, 171
rights *in* the city 24, 25, 36
right to inhabit the city 27
right to the city: basis in inequality analysis 28; collective power for 33–34; community activism to claim 33; Harvey's work on 25, 27; influence of 29–30; informational 26; lack of 25; Lefebvre's work on 26, 29–30; modest interventions 32–33; movement 34; as a movement 26; NUA agreement impact on 30; restricted 27; three-fold strategy for implementing 31–32; vs transcendent ideal of justice 27–28; *see also* ethical city
Rotorua, New Zealand 65, 66
Royal Society for the Encouragement of Arts, Manufactures and Commerce 135, 151
RSA *see* Royal Society for the Encouragement of Arts, Manufactures and Commerce
rubric 76
rule-based ethics *see* deontological ethics

rule consequentialism/utilitarianism 50
rules-based deontological approaches 63

San Diego, homeless ghettos in 51–52
Schwartz, Shalom 10
SDGs *see* Sustainable Development Goals
SDSN *see* Sustainable Development
    Solutions Network
self-governing benchmarking 61
sexual harassment, in Egypt 32
shared ethics 8, 56
sharing economies 54, 79, 177
Singapore 106
Singer, Peter 47, 49, 56
situational ethics 56
slum living 9
smart city technologies, ethical concerns
    related 136
social capital 4, 15
social change 9, 175
social contract 6, 18, 72 106, 115, 180
social economic arrangements 175, 177
social-economy-based enterprises 79
social housing and deontological ethics
    50–51
social ills, burden of responsibility for 44
Social Justice and the City (Harvey) 25
social mobilisation 33
social movements 7, 29, 33, 65
social self-referencing 56
social urban experiments 149
socio-economic ecosystems 79
sociology of innovation 170
socio-technical transitions 170–171
solar tax credit, in Hawaii 171
South Korea 132
spatial and class polarisation 108
Stanford prison experiment 104
Stockholm 133
street crime and unemployed youth, link
    between 119
Stronger Town Fund 146
subjective relativism 56
suffrage 24
super-rich '1 percent' class, rise of 56
surveillance capitalism 125
surveillance society 98
sustainability in urban policy, conditions
    for 28
Sustainable and Secure Buildings Act
    2004 98
Sustainable Development Goals 46, 111,
    157, 169, 177; approving and endorsing
    87; challenges in aligning policies with
    92; cost of supporting and measuring

89; critically important terms 88;
    goals 86, *87*; Local Agenda 21 (LA21)
    process 87–88, 90, 100; need for
    ethical engagement around 100–101;
    rationalisation 89; sustainability
    indicators for 88–89, **93**; top-down
    approach 88; 2015 UN resolution on
    88; voluntary local reviews (VLRs)
    89–92, *91*
Sustainable Development Solutions
    Network 89, 93
Sustainable Living Strategy (Rotorua) 65, 66

Te Arawa Partnership Model 65–66
technological sovereignty 136, 172
teleological approaches 63–64
teleology *see* consequentialism
Third Way 165, 167
top-down technocratic approach 101
Totnes, market town of 16, 17
transformative actions, imbalance
    between collective and individual 6
transition: initiatives 170–171;
    potential 179
transnational corporations, global
    hierarchies of 106
Transparency International 77, 99, 114
triple bottom line approach 89
Trump Administration 114
Turkey, development and housing
    policies in 32

UBI experiments *see* universal basic
    income experiments
UCLG *see* United Cities and Local
    Governments
UK, casualisation of work in 126
UN Global Compact Cities Programme
    66, 92
United Cities and Local Governments 89
United States: economic changes in 127;
    2016 elections, conspiracy stories
    around 76; employment exposed to
    automation 128–129; manufacturing
    jobs, decline in 126–127; ranked as
    second-tier country 127
universal basic income experiments
    54–55, 79, 137, 158, 165, 167, 179;
    B-MINCOME project 152, **153**;
    in Finland 151; legislating for 47; in
    Maricá, Brazil 153; in Ontario, Canada
    151–152; potential 153
universal rights 25, 168, 180
urban agglomeration 14
urban change 7, 125, 133, 168, 173, 178

urban communities and neoliberal project 44
urban crisis 42
urban design decision, ethical quandaries from 41
urban development, sustainable forms of 9
urban discourses 16
urban environmental justice, campaigns for 28
urban ethics 16–17, 28, 44
urban governance systems 27–28
urban growth 105, 133
urban high street 17
urban inequality 10, 13–14, 56; carbon intensity and income 99; and utilitarianism 50; wealth 8
urbanism 16; ethical approach to 2, 9–10; ethical dimensions of 15; Marx-informed analysis of 29
urban justice 15
urban liveability orientation see liveable city orientation
urban living labs 170
urban mega-slums 115
urban militarisation, public reaction to 114
urban planning 29; gender gap in 28; policies, neoliberalism impact on 67, **68–70**, 71; urban ethical frameworks 16
urban poverty 114, 157
urban projects 47
urban public spaces, privatisation of 31
urban renewal projects 32
urban security concerns 98
urban sustainability indicators 88–89
urban transitions: building alliances for 179–180; to ethical city, opportunities for promoting 172–173; initiatives 170–171; interventions and interruptions promoting 171–172; politics in 172; potential 179
Urban Transitions Alliance 179
urban uprising 119
utilitarianism 49–50
utopian cities 18–19, 124–126
utopian thinking 125

values, disconnect between cities and 10–14; basic values 10–11, *12*; city dwellers 10; climate change 13–14; in diverse cultures, surveys on 10–11; life goals and motivational values 11, *13*; physical aspects of 16; social inclusion 13–14; urban inequality 13–14
VCAT *see* Victorian Civil and Administrative Tribunal
Venice 18
Victorian Civil and Administrative Tribunal 52
Vienna 50, 110, 126
violence against women 28–29, 32
virtue ethics 52–53
virtuous cycle 3
VLRs *see* voluntary local reviews
voluntary local reviews 89–92, *91*, 157, 169; Bristol 90, 92; challenges associated with 92; characteristics of 90; ethical city assessment 98–100; in first cycle of 2030 Agenda *91*; institutional models for development of 90; pioneered by Japanese cities 89–90; reasons to undertake 90
vulnerable groups 13, 48, 64

We (Zamyatin) 124–125
wealth inequality 8, 136
Western Modernist planning theory 16
women: promoting actions by 32; resistance to Ankara Metropolitan Municipality 32–33; rights *to* the city 28–29, 32–33; violence against 28–29, 32
work scenarios, futures of *see* labour market, scenarios in 2035
World Charter on the Right to the City 29
world city system 106, 108

zero-carbon urban futures 132–135
Zimbardo, Philip 104
Zurich: carbon emissions reductions by 175; potential measures to meet 2050 targets *176*

Printed in the United States
By Bookmasters